D1224275

# Principles of
# Medical Imaging

# Principles of Medical Imaging

## K. Kirk Shung

Department of Bioengineering
Pennsylvania State University
University Park, Pennsylvania

## Michael B. Smith

Department of Radiology
Milton S. Hershey Medical Center
Pennsylvania State University
Hershey, Pennsylvania

## Benjamin M. W. Tsui

Departments of Biomedical Engineering and Radiology
University of North Carolina at Chapel Hill
Chapel Hill, North Carolina

**Academic Press, Inc.**
Harcourt Brace Jovanovich, Publishers

San Diego   New York   Boston   London   Sydney   Tokyo   Toronto

Academic Press, Inc.
1250 Sixth Avenue, San Diego, California 92101-4311

*United Kingdom Edition published by*
Academic Press Limited
24–28 Oval Road, London NW1 7DX

Library of Congress Cataloging-in-Publication Data

Shung, K. Kirk.
   Principles of medical imaging / K. Kirk Shung, Michael B. Smith,
Benjamin Tsui.
   Includes bibliographical references and index
      p.   cm.
   ISBN  0-12-640970-6
      1. Diagnostic imaging.   2. Medical physics.   I. Smith, Michael B.
(Michael Bruce), date.   II. Tsui, Benjamin.   III. Title.
   [DNLM: 1. Biophysics.   2. Diagnostic Imaging--instrumentation.
WN 200 S562p]
RC78.7.D53S52   1992
616.07'54--dc20
DNLM/DLC
for Library of Congress                                              92-7143
                                                                          CIP

PRINTED IN THE UNITED STATES OF AMERICA
92 93 94 95 96  QW  9 8 7 6 5 4 3 2 1

To Linda, Albert, Simon, and May Shung

To Thomas Smith, Jr. and Ruth Smith

To Amy, Andrew, and Kevin Tsui

# Contents

CHAPTER 2    **Ultrasound**

*K. Kirk Shung*

## CHAPTER 3   Radionuclide Imaging

*Benjamin M. W. Tsui and K. Kirk Shung*

CHAPTER 4  **Magnetic Resonance Imaging**

*Michael B. Smith, K. Kirk Shung, and Timothy J. Mosher*

# Preface

The field of medical imaging is growing at a rapid pace. Since the early 1960s, three new imaging modalities, namely, radionuclide imaging, ultrasound, and magnetic resonance imaging, have appeared and matured. Along with X-ray they are among the most important clinical diagnostic tools in medicine today. Radionuclide imaging, although its resolution cannot match that of other modalities, uses radioactive isotopes attached to biochemically active substances to yield unique information about the biochemical or physiological function of the organ which is unattainable otherwise. Ultrasound scanners use high frequency sound waves to interrogate the interior of the body. They are capable of depicting anatomical details with excellent resolution. Ultrasound is particulary suited to situations where exposure to ionizing radiation is undesirable, such as in obstetrical and neonatal scanning, and to imaging structures in motion, such as heart valves. Magnetic resonance imaging, however, has been envisioned to be the most exciting of them all by far because it also uses a form of nonionizing radiation, can achieve superior resolution, and is capable of yielding physiological information. In this period, significant progress has also been achieved in conventional X-ray radiography. Improved design or introduction of better materials in image intensifiers, intensifying and fluoroscopic screens, and photographic films has enhanced the resolution to a significant degree without adding higher patient radiation exposure levels. It is therefore plausible to understand why conventional radiography is still routinely used clinically for the diagnosis of many diseases and is the gold standard to which newer imaging modalities are compared.

Unquestionably, the digital revolution is the primary reason that has caused the medical imaging field to experience the explosive growth that we are seeing today. Computer and digital technology along with advances in electronics have made data acquisition fast and mass data storage possible. These are the most essential ingredients for the practical realization of tomographical reconstruction principles. X-ray computed tomography (CT), digital radiography, real-time ultrasonic scanners, single-photon emission computed tomography (SPECT), positron emission tomography (PET), and magnetic resonance imaging (MRI), which came about after the early 1970s, are just a few well-known products of the digital revolution in medical imaging.

While the development of these new imaging approaches may have contributed greatly to the improvement of health care, it has also contributed to the rising cost of health care. A chest X-ray costs only $20–30 per procedure whereas a magnetic resonance scan may cost up to $1000, let alone the expenses associated with acquiring and installing such a scanner. The cost-to-benefit ratio for these expensive procedures in certain cases is sometimes not as clear as in others. Therefore it is not unusual that the clinical efficacy and contribution of these modalities to patient care are being scrutinized and debated constantly by the medical community as well as the public.

This book is intended to be a university textbook for a senior or first-year graduate level course in medical imaging offered in a biomedical engineering, electrical engineering, medical physics, or radiological sciences department. Much of the material is calculus based. However, an attempt has been made to minimize mathematical derivation and to place more emphasis on physical concepts. A major part of this book was derived from notes used by the authors to teach a graduate course in medical imaging at the Bioengineering Program of Pennsylvania State University since the late 1970s. This book covers all four major medical imaging modalities, namely, X-ray including CT and digital radiography, ultrasound, radionuclide imaging including SPECT and PET, and magnetic resonance imaging. It is divided into four chapters in which a similar format is used. In each chapter fundamental physics involved in a modality is given first, followed by a discussion on instrumentation. Then various diagnostic procedures are described. Finally, recent developments and biological effects of each modality are discussed. At the end of each chapter a list of relevant references, further reading materials, and a set of problems are given. The purpose of this textbook is to give students with an adequate background in mathematics and physics an introduction to the field of diagnostic imaging; the materials discussed should be more than sufficient for one semester. However, the book may also be used as the text for a two-semester course in medical imaging when supplemented by additional materials or by inclusion of more mathematic detail.

Although this book has been written as a college textbook, radiologists with some technical background and practicing engineers or physicists working in imaging industries should also find it a valuable reference in the medical imaging field. As a final note, it should be pointed out that there are other imaging methods that have been used in medicine [e.g., thermography, magnetic imaging, and microwave imaging (Hendee, 1991)]. They are not included in this book primarily due to their limited utility at present. Readers who are interested in these modalities may refer to several books listed in the following reference section.

## References and Further Reading

Barrett, H., and Swindell, W. (1981). "Radiological Imaging: The Theory of Image Formation, Detection, and Processing." Academic Press, New York.

Curry, T. S. III, Dowdey, J. E., and Murry, R. C. Jr. (1990). "Christensen's Introduction to the Physics of Diagnostic Radiology, 4th Edition." Lea and Febiger, Philadelphia.

Freeman, S. E., Fukushima, E., and Greene, E. R. (1990), "Noninvasive Techniques in Biology and Medicine." San Francisco Press, San Francisco.

Hendee, W. R. (1991). *Diagnostic Imaging* **13**, A13.

Huang, H. K. (1987). "Elements of Digital Radiology." Prentice-Hall, Englewood Cliffs, New Jersey.

Jaffe, C. C. (1982). *Am. Scient.* **70**, 576.

Krestel, E. (1990). "Imaging Systems for Medical Diagnosis." Siemens, Aktiengesellschaft, Germany.

Larsen, L. E., and Jacobi, J. H. (1985). "Medical Applications of Microwave Imaging." IEEE Press, New York.

Mackay, R. S. (1984). "Medical Images and Display." John Wiley, New York.

Macovski, A. (1983). "Medical Imaging Sysems." Prentice-Hall, Englewood Cliffs, New Jersey.

Nudelman, S., and Patton, D. D. (1980). "Imaging for Medicine, Vol. 1." Plenum Press, New York.

Sochurek, H. (1987). *National Geographic,* **171,** 1.

Swenburg, C. E., and Conklin, J. J. (1989). "Imaging Techniques in Biology and Medicine." Academic Press, San Diego.

Wallace, J. D., and Cade, C. M. (1975). "Clinical Thermography." CRC Press, Boca Raton, Florida.

Webb, S. (1988). "The Physics of Medical Imaging." Adam Hilger, Bristol and Philadelphia.

Wells, P. N. T. (1982). "Scientific Basis of Medical Imaging." Churchill Livingston, Edinburgh and New York.

# Acknowledgments

A large part of this book is based on the notes developed by one of the authors (KKS) for teaching a three-credit, senior to graduate level course on medical imaging in the Bioengineering Program at Pennsylvania State University.

The authors would like to express their thanks to Mrs. Rita Kline and Ms. Brenda Bixler for preparing and typing part of the manuscript; the editors Mr. Chuck Arthur and Ms. Aimee Squires at Academic Press for their assistance in making this book a reality; several current and former students for proofreading the manuscripts and for their valuable comments; and Dr. Timothy J. Mosher for contributing to Chapter 4.

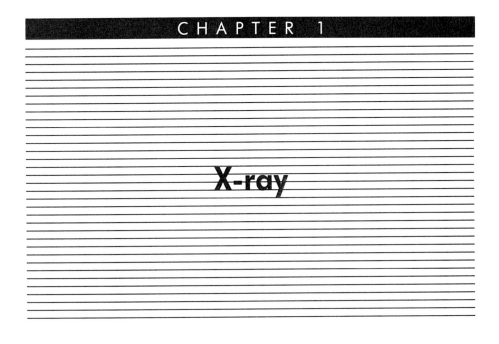

# CHAPTER 1

# X-ray

# I. Fundamentals of X-ray

X-ray was discovered accidentally by a German physicist, W. K. Roentgen, in 1895. Because of this discovery, he was awarded the first Nobel Prize in 1901. Since then, it has been the most important and most widely used tool in clinical medicine for almost a century. X-ray is a form of electromagnetic (EM) energy just like radio waves or light. However, there are also significant differences.

## A. Electromagnetic Radiation

Table I shows the EM radiation spectrum. X-ray, light, radio waves, and microwaves are all well-known members of the EM spectrum. EM radiation propagates as a wave characterized by a number of wave parameters, which are electric field, magnetic field, dielectric constant, and other electromagnetic properties of the medium. Assuming a plane monochromatic radiation, the magnetic and electric fields are both functions of time and space and can be represented by the following equation:

$$\phi(x, t) = \phi_0 \cos(\omega t - kx) \qquad (1\text{-}1)$$

where $\phi$ is the electric field, $x$ is the distance that the wave has traveled, $t$ is time, and $\omega$ is the angular frequency. $\omega = 2\pi f$ where $f$ is the frequency and $k$ is the wave number given by $2\pi/\lambda$ where $\lambda$ is the wavelength. This equation is graphically represented in Fig. 1. Figure 1(a) shows the wave as a function of time at a fixed $x$ and Fig. 1(b) shows the wave as a function of $x$ at a fixed time. The wavelength is defined as the distance over which one cycle occurs. The period $T$ is the

## Table I
Electromagnetic wave spectrum

| Energy (eV) | Frequency (Hz) | | Wavelength (m) |
|---|---|---|---|
| $4 \times 10^{-11}$ | $10^4$ | | $10^4$ |
| $4 \times 10^{-10}$ | $10^5$ | AM radio waves | $10^3$ |
| $4 \times 10^{-9}$ | $10^6$ | | $10^2$ |
| $4 \times 10^{-8}$ | $10^7$ | Short radio waves | $10^1$ |
| | | FM radio waves and TV | |
| $4 \times 10^{-7}$ | $10^8$ | | $10^0$ |
| $4 \times 10^{-6}$ | $10^9$ | | $10^{-1}$ |
| $4 \times 10^{-5}$ | $10^{10}$ | Microwaves and radar | $10^{-2}$ |
| $4 \times 10^{-4}$ | $10^{11}$ | | $10^{-3}$ |
| $4 \times 10^{-3}$ | $10^{12}$ | Infrared light | $10^{-4}$ |
| $4 \times 10^{-2}$ | $10^{13}$ | | $10^{-5}$ |
| $4 \times 10^{-1}$ | $10^{14}$ | Visible light | $10^{-6}$ |
| $4 \times 10^{0}$ | $10^{15}$ | Ultraviolet light | $10^{-7}$ |
| $4 \times 10^{1}$ | $10^{16}$ | | $10^{-8}$ |
| $4 \times 10^{2}$ | $10^{17}$ | | $10^{-9}$ |
| $4 \times 10^{3}$ | $10^{18}$ | X-ray | $10^{-10}$ |
| $4 \times 10^{4}$ | $10^{19}$ | | $10^{-11}$ |
| $4 \times 10^{5}$ | $10^{20}$ | | $10^{-12}$ |
| $4 \times 10^{6}$ | $10^{21}$ | Gamma ray | $10^{-13}$ |
| $4 \times 10^{7}$ | $10^{22}$ | Cosmic ray | $10^{-14}$ |

time that it takes for one cycle to occur. It is then quite obvious that

$$Tc = \lambda$$

where $c$ is the speed of electromagnetic wave propagation. In vacuum, $c$ is $3 \times 10^8$ m/sec. Since $f = 1/T$, we have

$$f\lambda = c \qquad (1\text{-}2)$$

Equation (1-1) is a solution to the wave equation

$$\frac{\partial^2 \phi}{\partial x^2} = \frac{1}{c^2} \frac{\partial^2 \phi}{\partial t^2} \qquad (1\text{-}3)$$

where $c = (1/\mu\epsilon)^{-1/2}$, $\mu$ is permeability of the medium and equals $1.257 \times 10^{-6}$ henry/m in free space, and $\epsilon$ is permittivity of the medium and equals $8.854 \times 10^{-12}$ Farad/m in free space. The general solution to the wave equation is of the form $f(t \pm x/c)$, where the negative sign denotes a wave traveling in the $+x$ direction whereas the positive sign indicates a wave traveling in the $-x$ direction (Hirose and Lonngren, 1985). Equation (1-1) is a special form of the solution to the wave equation.

The main difference between X-ray and light or radio waves is in their frequency or wavelength. Diagnostic X-rays typically have a wavelength from 100 nm to 0.01 nm, which is much shorter.

Unlike light or radio waves, the propagation of X-ray radiation is sometimes difficult to interpret by treating the radiation as a wave alone. It is sometimes necessary to discuss X-ray radiation as if it possessed the dual characteristics of both particles and waves.

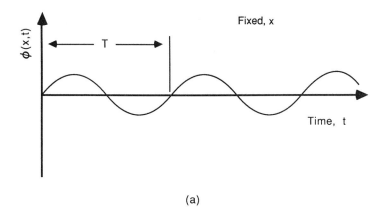

(a)

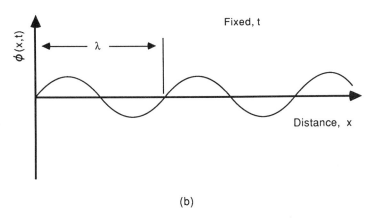

(b)

**Figure 1**  Sinusoidal electromagnetic wave (a) as a function of time at fixed distance and (b) as a function of distance at a certain instant of time.

The wave concept is useful in explaining such phenomena as the reflection, scattering, refraction, and diffraction characteristics of X-rays. However, in a number of situations, the particle concept has to be used.

In discussing its particle properties, the X-ray radiation is thought to be particles traveling at the speed of light and carrying an energy given by $E = hf$, where $h$ is the Planck constant ($4.13 \times 10^{-18}$ keV-sec where 1 eV = $1.6 \times 10^{-19}$ joules). These particles are called quanta, or photons. A photon having an energy level greater than a few electron volts is capable of ionizing atoms and molecules (i.e., knocking an electron out of its orbit), and it is called ionization radiation.

Consider an X-ray photon with a wavelength of 1 nm. The energy of the photon is then

$$E = 4.13 \times 10^{-15} \times 3 \times 10^8/10^{-9} = 1.2 \times 10^3 \text{ eV}$$

Therefore, X-ray is an ionizing radiation. It can be easily shown that gamma rays and some ultraviolet rays are ionizing radiation as well.

## B.    Interactions between X-rays and Matter

X-rays can interact with the orbital electrons as well as with the nuclei of the atoms. In the diagnostic energy range, the interactions are more likely to involve orbital electrons, because their energy levels are too low to interact with nuclei. There are five ways that X-ray photons can interact with atoms or molecules in matter. They are (1) coherent scattering, (2) photoelectric effect, (3) Compton scattering, (4) pair production, and (5) photodisintegration. The probability for each process to occur depends on the energy of the X-ray photon and the atomic number of the atom among other things.

### 1.    Coherent Scattering

When a photon collides with another particle, the photon is deflected into another direction losing little energy and, therefore, with negligible change in wavelength. This type of interaction is called coherent scattering (Fig. 2). Coherent scattering generally occurs in low-energy radiation that does not carry enough energy to eject the orbital electrons out of the orbit or ionize the atom or molecule. This is the only interaction of the five between X-ray and matter that does not cause ionization.

### 2.    Photoelectric Effect

The atom can be considered as an energy well with the nucleus situated at the bottom of the well as far as the electrons are concerned. The electrons closer to the nucleus require more energy to escape the well or to become a free electron while the electrons farther from the nucleus require less energy. The depth of the energy well is determined by the atomic number.

Consider the example of iodine, which has an atomic number of 53. The K-shell, L-shell, M-shell electrons, respectively, have binding energy levels of $-33.2$, $-4.3$, and $-0.6$ keV. Therefore, for a K-shell electron to escape the nucleus, 33.2 keV energy is required. On the other hand, for a K-shell electron of lead to become a free electron, an energy of 88 keV is needed.

The photoelectric effect is graphically illustrated in Fig. 3. An incident X-ray photon, which carries energy slightly higher than the binding energy of a K-shell

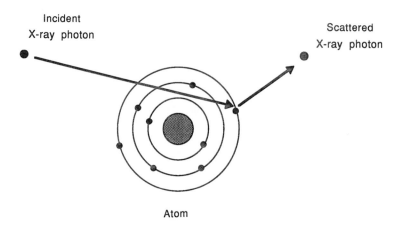

**Figure 2**   Coherent scattering of an X-ray photon by an atom.

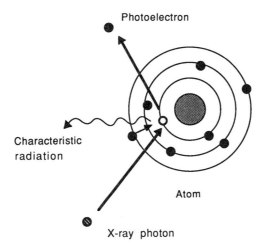

**Figure 3**  Photoelectric effect.

electron, collides with one of these orbital electrons and ejects it from the orbit. The photon is said to be absorbed, giving up all its energy. The majority of the photon's energy is needed to overcome the binding energy and the rest gives the escaped electron kinetic energy. This freed electron is called a photoelectron. The vacancy in the orbit generated due to the escaped electron will be filled almost instantly, because the most stable state of an atom is its lowest energy state, by an electron from outer shells. The remaining atom now becomes a positively charged ion. Accompanying the ionization, a characteristic radiation or fluorescent radiation in the form of an X-ray photon will be emitted carrying an energy equal to the difference in energy between the outer shell electron and the K-shell electron. As an alternative to characteristic radiation, the Auger effect, in which an outer shell electron fills the inner shell vacancy but the energy released by the outer shell electron is transferred to another orbital electron instead of being emitted as characteristic radiation, may occur. The orbital electron that acquires enough energy to escape is called an Auger electron. The two orbital vacancies after the emission of an Auger electron are filled by other outer shell electrons, resulting in more characteristic radiation or Auger electrons. The probability for characteristic radiation to occur, defined as fluorescent yield, is a function of the atomic number of the atom. Heavier atoms are more likely to emit characteristic radiation whereas lighter atoms are more likely to emit Auger electrons. Thus, the photoelectric effect always yields three end products: (1) characteristic radiation or Auger electrons, (2) a negative ion (photoelectron), and (3) a positive ion. For iodine the characteristic radiation may occur at 33.2, 32.6, or 28.3 keV if a K-shell electron is ejected from the orbit, depending on whether an N, M, or L-shell electron falls into the void.

The probability for the photoelectric effect to occur is governed by the following principles that can be deduced from quantum mechanics. (1) The energy carried by the incident X-ray photon has to be higher than the binding energy of the orbital electron for the electron to be ejected. (2) The probability for the photoelectric effect to occur is approximately proportional to the third power of the atomic number of the element and inversely proportional to the third power of the photon energy once the photon energy becomes greater than the binding energy of the electron (Hobbie, 1978). Therefore, when the probability of photoelectric effect to occur for an element is plotted as a function of photon energy, sharp maxima always occur at the binding energies of orbital electrons. This matter will be discussed again later in this chapter.

Photoelectric effect is the most desirable type of interaction in X-ray imaging because the X-ray photon is completely absorbed, producing little scattered radiation. The scattered radiation is a health hazard to the personnel using the equipment and is a form of image noise, which degrades the image quality.

## 3.   Compton Scattering

The scattered radiation encountered in an X-ray examination almost exclusively results from Compton scattering. Compton scattering is different from photoelectric effect in that only part of the energy carried by the photon is transmitted to the electron; in other words, the photon is scattered by the electron into other directions with a reduction in energy or an increase in wavelength whereas in photoelectric effect as defined in diagnostic radiology the X-ray photon is completely absorbed by the atom. The Compton scattering is graphically portrayed in Fig. 4. The photon collides with an outer-shell electron and knocks it out from its orbit. The photon is deflected by the electron into other directions, retaining some of its energy. The amount of energy that a scattered photon may retain depends on two factors: the initial photon energy relative to the binding energy of the orbital electron and the scattering angle ($\theta$ in Fig. 4) according to the following equation under the assumption that the electron is free and stationary before collision.

$$E' = \frac{E}{1 + \left(\dfrac{E}{m_e c^2}\right)(1 - \cos\theta)}$$

where $E$, $E'$, and $m_e$ are respectively the energy of the incident photon, the energy of the scattered photon, and the rest mass of the electron equivalent to 511 keV of energy. If photon energy is relatively low, the energy carried by the scattered photon is almost independent of the angle. It can be readily seen that this is the case of isotropic scattering in considering wave scattering phenomenon where wavelength is much larger than size of the scatterer. As the energy is in-

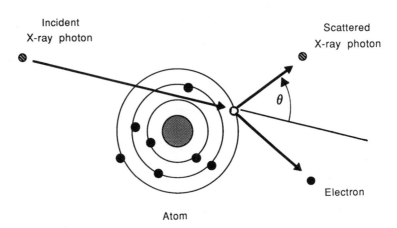

Figure 4   Compton scattering of an X-ray photon by an atom.

creased, photons scattered at small angles or in the forward direction carry higher energy. An exact analogy again can be found in wave scattering: Scattering is concentrated in the forward direction for wavelength much smaller than the scatterer size.

Compton scattering creates two major problems in X-ray radiography. First, it produces the so-called background noise on the film and, second, it is a major safety hazard for the personnel using the equipment because the scattered photons possess large amounts of energy and can escape the patient.

### 4.  Pair Production and Photodisintegration

In these interactions, the nucleus is involved. Generally, they have little importance in X-ray radiography because very high energy photons in the order of 1 MeV are required for these interactions to occur. Therefore, they will be described only briefly here but will be discussed in more detail in Chapter 3 on nuclear imaging.

Pair production pertains to the process in which the high-energy photon is completely absorbed by the nucleus and converted into two particles: an electron and a positron, a particle with the same mass as an electron but with a positive charge. Photodisintegration on the other hand describes a process in which one or more nuclear particles or nucleons such as neutrons or protons are ejected from the nucleus by the high-energy photon.

In summation, three different interactions can occur as an X-ray photon in the diagnostic range encounters an atom. Which interaction will occur depends on the energy of the photon and the binding energies of the electrons. Photoelectric effect is most likely to occur when the photon energy matches the binding energies.

## C.  Intensity of an X-ray Beam

The intensity of a beam is defined as the power per unit area of the beam. The power contained in an X-ray beam is related to the number of photons crossing the beam cross-sectional area per unit time and the energy carried by these photons. Consequently, the intensity of an X-ray beam is a function of both the number of photons and the energy of the photons. Therefore, the intensity of the X-ray beam can be changed by either varying the number of photons or varying the energy of the photon.

In X-ray, two units are frequently used to indicate the energy associated with an X-ray beam.

### 1.  Roentgen (R)

The reason that X-ray or nuclear radiation is called ionizing radiation is that it possesses enough energy to cause the ionization of atoms. The unit roentgen is a measure of X-ray energy that is the product of power contained in the beam and time of exposure and is defined as the total number of ion pairs produced in 1 cc of air under standard conditions (at 760 mm Hg ambient pressure and 0°C) by the radiation or the amount of radiation that produces an electric charge separation of $2.58 \times 10^{-4}$ coulomb per kg of air.

### 2.  Radiation Absorbed Dose (rad)

This unit defines the amount of radiation actually absorbed by a medium. A rad means that 0.01 joule of energy is absorbed by 1 kilogram of material. The reason for introducing this additional unit is that different materials have different X-ray absorption characteristics. The amount of energy absorbed by different materials may be different for the same amount of radiation. This difference depends on the absorption characteristics of the material and X-ray photon energy, as will become clear in the next section. Another unit, gray (Gy) is also used and denotes 100 rad.

## D.  Attenuation

As a beam of X-ray is incident upon a block of material, the X-ray intensity decreases because of the interaction of the photons and the material as described. Assume that the X-ray beam has an intensity $I$ and a cross-sectional area $A$. Also assume that the atoms in the material are identical and all have a cross section of $\sigma$ and there are $n$ atoms per unit volume of the material. Then, as shown in Fig. 5, the total number of atoms encountered by the X-ray beam is given by $An$ and the area occupied by the atoms in the beam is $An\sigma$. Thus the probability for a photon to interact with an atom is $An\sigma/A = n\sigma$. Therefore, the X-ray energy removed in thickness $dx$ is

$$dI = -n\sigma I \, dx$$

Rearranging the equation, we have

$$dI/dx = -n\sigma I$$

Let $\beta = n\sigma$ which is the fraction of X-ray energy removed per unit thickness per unit intensity. Substituting $\beta$ into the above equation and carrying out the integration, we obtain

$$I = I_0 e^{-\beta x} \tag{1-4}$$

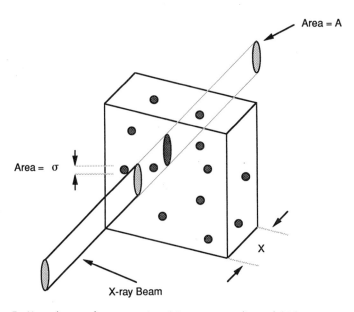

**Figure 5**   X-ray beam of cross-section $A$ intersects medium of thickness $x$.

where $I$ is X-ray intensity at $x$, $I_0$ is incident X-ray intensity, $\beta$ is defined as the linear attenuation coefficient in np/cm or cm$^{-1}$, and $x$ is the propagation length.

Therefore, the propagation length required to reduce to the intensity of the original beam by $1/2$ is given by

$$\text{HVL (Half-value layer)} = \frac{0.693}{\beta} \qquad (1\text{-}5)$$

Note here that the linear-attenuation coefficient and, thus, HVL are the function of photon energy. The mass-attenuation coefficient is defined as the ratio of linear-attenuation coefficient to density, or $\beta/\eta$ where $\eta$ represents the mass density of the material. The unit of the mass-attenuation coefficient is cm$^2$/gm. The reason for introducing the mass coefficient is that we would like to have a quantity indicating the attenuation property of the matter independent of its physical state. For instance the linear-attenuation coefficients for water, ice, and water vapor at 50 keV are respectively 0.214, 0.196, and 0.00013 cm$^{-1}$. However, their mass-attenuation coefficients are the same, 0.214 cm$^2$/gm.

The mass-attenuation coefficients of fat, bone, muscle, iodine, and lead are plotted as functions of energy in Fig. 6.

## 1.  Factors That Affect Attenuation Coefficients

Although X-ray photons can interact with both the nucleus and electrons of an atom, photoelectric effect and Compton scattering are the most dominant mechanisms for attenuation in the diagnostic X-ray energy range. Coherent scattering

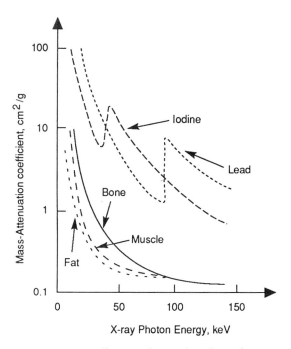

**Figure 6**  Mass-attenuation coefficients of several media as functions of X-ray energy.

usually is negligible. When X-ray photons interact with atoms in a material, the X-ray photons may be completely or partially scattered giving up some energy, or completely absorbed. In other words, two mechanisms contribute to the attenuation of an X-ray beam: absorption and scattering. Therefore,

$$\beta = \beta_s + \beta_a$$

where $\beta_s$ and $\beta_a$ are respectively the fractions of attenuation due to scattering and absorption. $\beta_a$ is called X-ray absorption coefficient. Similarly $\beta_a/\eta$ is defined as the mass absorption coefficient.

What affect the attenuation of a material and the relative importance of scattering and absorption in attenuation have depends on several factors among which are photon energy, atomic number, density, electron density (number of electrons per unit mass), and thickness of the material (Wilkes, 1981).

Attenuation due to coherent or elastic scattering, $\beta_{coh}$, which occurs when X-ray photon energy is smaller than the binding energies of orbital electrons, is related to atomic number, $Z$, density, $\eta$, and photon energy, $E$, by the following equation.

$$\beta_{coh} \approx \eta Z^2 E^{-1}$$

In coherent scattering, the scattered photons retain most of their energy and there is little energy being absorbed. Consequently, attenuation due to coherent scattering phenomenon is caused mainly by scattering. Further, since the average atomic numbers of biological tissues are low and X-ray photon energy in the diagnostic range is relatively high, the contribution of coherent scattering to the overall X-ray attenuation of a material is negligible.

Attenuations due to photoelectric effect, $\beta_{pho}$, and Compton scattering, $\beta_{com}$, are related to density of the material and photon energy respectively by

$$\beta_{pho} \approx \eta Z^3 E^{-3}$$
$$\beta_{com} \approx \eta \rho_e E^{-1}$$

where $\rho_e$ is the electron density with a unit (number of electrons)/gm. The electron density is related to the atomic number $Z$ by $\rho_e = N_{av} Z/A_m$ where $N_{av}$ is Avogadro's number, $6.023 \times 10^{23}$, and $A_m$ is the atomic mass. Because $A_m$ is the sum of the number of protons and the number of neutrons in the nucleus and $Z$ equals the number of protons, $\rho_e$ is inversely related to the number of neutrons in the nucleus. The hydrogen nucleus has no neutron. Thus, $\rho_e$ for hydrogen is the largest, being $6.023 \times 10^{23}$ electrons/gm, and $\rho_e$ for all other elements is in a narrow range from 2.5 to $3.5 \times 10^{23}$ electrons/gm. As opposed to coherent scattering, attenuation due to photoelectric effect is dominated by absorption whereas the relative contribution of absorption and scattering to attenuation for a substance is a complicated function of energy in Compton scattering process.

Since

$$\beta = \beta_{coh} + \beta_{pho} + \beta_{com'}$$

it is clear that X-ray attenuation is linearly proportional to the density of the material and is related to atomic number and electron density in a more complex way.

It is also clear from these relationships that attenuation decreases as photon energy increases. This behavior can be easily understood by treating the X-ray photons as particles whose size is represented by the wavelength. As photon energy increases, the wavelength becomes smaller, or in other words, the particles become smaller. It is quite obvious that the smaller the particles, the easier for them to traverse the matrix of electrons and nuclei. Although this rule is generally true, there is an exception. Recall that the attenuation of an X-ray beam is caused by three factors: coherent scattering, photoelectric effect, and Compton scattering. The probability for the photoelectric effect to occur is maximal when the photon energy approaches the binding energies of orbital electrons. As a result, when the radiation energy level approaches the binding energies of orbital electrons, there will be an abrupt increase in attenuation. As the energy level is further increased, the attenuation decreases. Figure 6 shows the mass-attenuation coefficient for lead and iodine over the diagnostic energy level. From this figure, it becomes apparent that iodine is a contrast medium in diagnostic radiology, because its K-edge is at 33.2 keV, or it is very absorbent at 33.2 keV.

It is important to note here that Eq. (1-4) applies only to monochromatic radiation. X-ray photons generated by X-ray sources typically are heterogeneous or polychromatic that is, distributed over a range of energies. Since photons of different energy are attenuated differently after propagating through a medium, as was discussed, the transmitted intensity of a polychromatic or heterogeneous X-ray beam can not be calculated using Eq. (1-4). In practice, the energy of a monochromatic beam that has the same half-value layer as the polychromatic beam in question (as in a medium such as aluminum) is considered the effective energy of the polychromatic beam (Bange, 1988). The polychromatic beam after traversing a medium contains less photons in the lower energy range, causing the effective energy of the beam to increase. This phenomenon is called "beam hardening."

# II.  Generation and Detection of X-rays

## A.  X-ray Generation

X-rays are generated when electrons with high energy strike a target made from materials like tungsten or molybdenum. The high-energy electrons can interact with the nuclei of the tungsten atoms producing the so-called general radiation or white radiation or Bremsstrahlung, or they can interact with the orbital electrons producing the characteristic radiation.

### 1.  White Radiation

When an electron that is negatively charged passes near the positively charged nucleus, the electron is attracted toward the nucleus and then deflected from its original path. The electron may lose some energy or may not. If it does not, the process is called elastic scattering and no X-ray photons will be produced. If it does lose energy, the process is called inelastic scattering and the energy lost by the electron is emitted in the form of an X-ray photon. The radiation produced in this way is called white radiation, which is depicted in Fig. 7. The probability of the electron to lose energy is increased as the atomic number of the atom increases. The electrons striking the target can interact with a number of nuclei before being stopped and the electrons may carry different energies. Therefore, the energies of the X-ray photons generated by the process of general radiation are distributed over a wide range, as shown in Fig. 8.

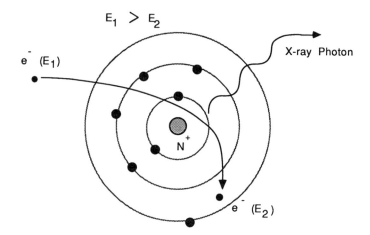

**Figure 7** Deflection of high-energy electron by nucleus produces white radiation.

## 2. Characteristic Radiation

When the electrons striking the target interact with orbital electrons in inner shells, characteristic radiation results. This process is very similar to that described in photoelectric effect. Figure 8 shows the typical X-ray spectrum generated at a tungsten target. The dashed line represents the general radiation observed in vacuum within the tube glass enclosure. The solid line represents the general radiation actually obtained outside the tube. $\alpha_1$, $\alpha_2$, $\beta_1$, and $\beta_2$ represent the characteristic radiations resulting from L shells falling into K shells (59.3 and 57.9 keV) and M and N-shell electrons falling into K shells (67.2 and 69 keV), respectively.

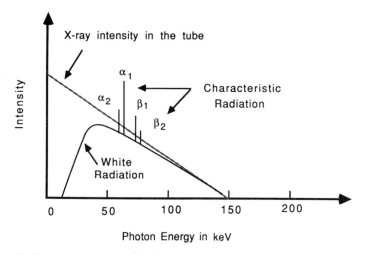

**Figure 8** X-ray spectrum produced by the tungsten target of an X-ray tube.

## B.  X-ray Generators

The basic construction of an X-ray tube is illustrated in Fig. 9. The electrodes are generally sealed in a vacuum, which allows independent control of the number and speed of the accelerated electrons striking the tungsten anode. The presence of gas can result in variation in the number of electrons and reduced speed. The cathode is composed of two elements, the filament, made of tungsten, and a metallic cup for focusing the electrons emitted by the filament. The filament is a helical coil of tungsten wire of about 0.2 mm in diameter. The coil is approximately 0.2 cm in diameter and 1 cm in length. When the current is fed through the wire, it becomes heated. The heat will be absorbed by the electrons in the wire. When the temperature reaches a certain level, the electrons absorb enough energy to overcome the surface barrier and to escape from the metal. These escaped electrons form a cloud around the filament and are called space charge. They prevent the electrons within the wire from escaping the filament, and the effect of the space charge on limiting the emission of more electrons from the filament is called the space charge effect. The electrons stay around the filament because the loss of electrons causes the filament to become positive. These escaped electrons can then be accelerated toward the anode by applying a high-voltage potential. Tungsten is desirable in X-ray application because it has a high melting point (3370°C) and little tendency to vaporize, and it is strong.

### 1.  Line Focus Principle

Most of the energy carried by the electrons bombarding the tungsten target on the anode is converted into heat (in fact, 99%). Therefore, a large focal spot is preferred because it allows the accumulation of larger amounts of heat. However, a small focal spot is needed to generate better images. This problem can be overcome by using the line focus principle as illustrated in Fig. 9. The anode angle, which is the angle between the slanted target surface and the plane perpendicular to the electron beam, typically varies from 5° to 15°. The effective focal size, $f$, is related to the length of actual focal size, $F$, on the anode by the following equation:

$$f = F \sin \theta$$

where $\theta$ is the anode angle. It is evident that a larger anode angle provides a larger area for bombardment but it also produces a larger apparent focal spot. In

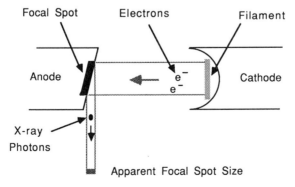

**Figure 9**   Basic components of an X-ray tube.

practice, the angle is limited by the so-called Heel effect, which is illustrated in Fig. 10. The intensity leaving the X-ray tube is not uniform; it is smaller in the anode direction because photons travelling in certain directions have to travel a longer path in the anode than others. The heat problem at the anode can be further reduced by using a rotating anode (3,000 to 10,000 rpm), which increases the total target area. A photograph of a modern X-ray tube is shown in Fig. 11.

## 2.   X-ray Tube Ratings

A number of factors can affect the intensity of the X-ray beam produced by the generator, namely: filament temperature controlled by the filament current ($i_f$), the potential difference between the anode and cathode (tube voltage, $V_t$), the number of electrons bombarding the anode target (tube current, mA), and the target material. The basic electric circuits involved in an X-ray generator are shown in Fig. 12 where $V_f$ is the source for providing the filament current.

**Target Material**   The higher the atomic number, the greater the efficiency of X-ray production. For example, platinum (atomic number 78) produces more white radiation than tungsten (atomic number 74) at the same tube current and potential.

**Tube Voltage**   The tube voltage $V_t$ can be either dc or ac following full-wave or half-wave rectification. For ac generators, it is usually measured in terms of peak voltage applied or kilovolts peak (kVp). The intensity is proportional to the square of kVp. The maximum energy produced also depends on the voltage. Typically the tube voltage ranges from a few kilovolts peak up to 150 kVp.

**Tube Current**   The number of X-ray photons produced depends on the number of electrons striking the target and therefore should depend on the tube current. It was found that the intensity is linearly proportional to the tube current. Typically the tube current ranges from a few milliamperes to a few hundred milliamperes.

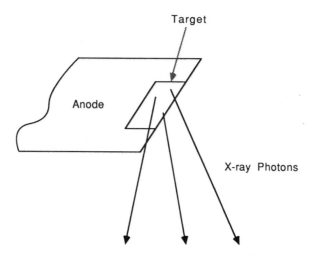

**Figure 10**   Heel effect that occurs in an X-ray tube produces a non-uniform X-ray field.

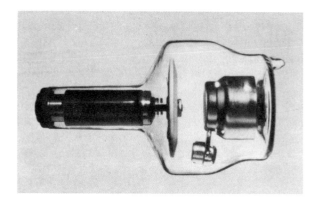

**Figure 11**   Photograph of X-ray tube.

**Filament Current**   The tube current increases initially as the tube voltage is increased at a fixed filament current. However, as the potential difference is further increased, a point will be reached after which an increase in potential difference has no effect on tube current. This point is called the saturation voltage. In this region, the current is limited by the filament temperature or the filament current. Typical values of filament current are a few amperes which can be dc or ac.

These observations can be summarized by the following equation at fixed $i_f$:

$$I \approx Z(\text{mA})(\text{kVp})^2 F$$

where $I$ is the intensity irradiated by the X-ray tube, $Z$ is the atomic number of the target material, and $F$ is the rectification factor for $V_t$ and is one for direct current. It is worthwhile to note here that a change in kVp results in a change in $E_{\max}$, the maximal energy that can be carried by an X-ray photon, whereas a change in mA does not.

## C.  Filters

Filtration in X-ray describes a process similar to that in electrical signal processing, that is, a process of removing signals of undesired frequencies from the input

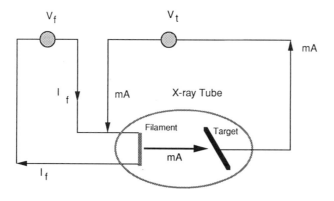

**Figure 12**   Electrical circuits associated with an X-ray generator.

signal. As was mentioned, X-rays generated by X-ray tubes are polychromatic. Depending on the nature of a certain application, only a portion of the energy spectrum is desirable. Therefore, the radiation dose to the patient can be substantially reduced by filtering out the undesired portion of the X-ray spectrum.

The X-ray photons emitted by the anode of an X-ray generator during an X-ray examination are absorbed by the X-ray tube itself, the filters or absorbers, and the patient before reaching the X-ray detector such as a photographic film. The added absorbers usually are thin sheets of metal placed between the X-ray source and the patient. Aluminum is an excellent absorber for low-energy X-rays, while copper is most useful for high-energy X-rays. In practice, copper is never used alone as an absorber. It is always used in combination with aluminum as a compound filter because the characteristic radiation produced by copper is about 8 keV, which is energetic enough to reach the patient and increase the skin dose. Thus, the aluminum layer is placed below the copper layer to absorb this radiation. An aluminum layer 3 mm thick can attenuate more than 90% of the X-ray energy at 20 keV. The non-uniform nature of the X-ray beam produced by X-ray generators may be compensated by wedge filters which are thinner on the anode side than on the other.

## D. Beam Restrictors and Grids

There are three types of X-ray beam restrictors: aperture diaphragms, cones and cylinders, and collimators. The basic function of a beam restrictor is to regulate the size and shape of the beam. A closely collimated beam can reduce patient exposure and generate less scatter radiation. The aperture diaphragm is illustrated in Fig. 13. It is basically a sheet of lead with a hole in the center whose size and shape determines those of the X-ray beam. The width of the penumbra $P$ is related to the source diameter by the following equation:

$$P = \frac{D}{L}I \qquad (1\text{-}6)$$

where $D$ is the width of the source, $I$ and $L$ are the distances between the source and the object, and between the object and the detector, respectively. Therefore, to reduce the penumbra, the source should be made as small as possible and the diaphragm should be positioned as far away from the source as possible.

Cones and cylinders are sometimes also used as beam restrictors, but they, along with the diaphragm, suffer from a major drawback: Only a limited number of beam sizes can be obtained. The collimator is the most popular beam restrictor for two reasons: The X-ray field size is adjustable, and a light beam can be used to indicate the exact size of the field. The physical configuration is illustrated in Fig. 14. The X-ray beam size is adjusted by the movable aperture diaphragm. The X-ray field is illuminated by a light beam from a light bulb located within the collimator the same distance from the center of the mirror as the X-ray source.

Scattered X-rays are noise that degrades image quality and increases patient exposure and therefore should be minimized. The most effective way of removing scatter radiation is the radiographic grid shown in Fig. 15. The grid is composed of a series of lead foil strips separated by X-ray transparent spacers which are either aluminum or organic material. The grid blocks the scattered radiation while letting the primary radiation pass. The grid ratio in Fig. 15 is defined as $h/g$, that is, the ratio between the height of the lead strips and the width of the gap between

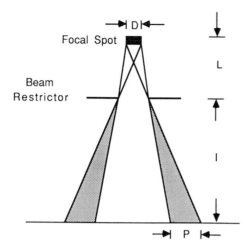

**Figure 13**   Effect of a beam restrictor on X-ray is illustrated. Finite aperture size results in penumbra along edges.

them. This ratio usually ranges from 4 to 16. For a grid with a height of 2 mm, lead strip width of 0.05 mm, and strip gap width of 0.25 mm, the grid ratio is 8. It then becomes quite apparent that the higher the grid ratio, the better the grid function in removing scatter radiation. However, here it has to be remembered that when a grid is used, we pay the price of increasing patient exposure if the X-ray intensity immediately behind the grid is to be kept the same as that prior to the insertion of the grid.

The grid shown in Fig. 15 is called a linear grid. Other forms of grids have also been used. When the grid strips are focused toward the X-ray source, the grid is called a focused grid.

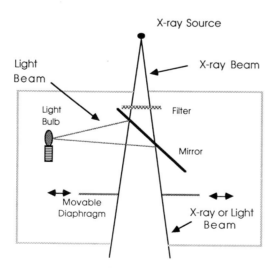

**Figure 14**   Physical construction of a collimator.

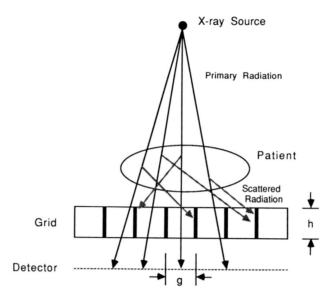

**Figure 15**  Scattered X-ray photons can be removed by positioning grid between patient and X-ray detector such as film.

If the lead strips are thin enough, their image on the film or the detector may be negligible. However, if the image quality requirement necessitates thick lead strips, the grid may be moved during exposure to blur out the image of the grid lines.

## E.  Intensifying Screens, Fluorescent Screens, and Image Intensifiers

The human eye cannot see the information carried by the X-rays directly. Therefore, the X-ray images have to be converted to visualizable information before they can be comprehensible to a human. In conventional radiography, this can be done either by exposing a photographic film to the X-ray or by converting the X-ray photons to visible photons. Because the sensitivity of the film to X-ray is very low, it is desirable to first convert the X-ray to visible photons by an intensifying screen, which can significantly reduce the patient dose while achieving a properly exposed film. The reduction of exposure time becomes important when it is necessary to minimize the effect of patient motion. The performance of an intensifying screen is maximized for emitting light photons at wavelengths that are optimal for exposing photographic films. Since the process of converting X-ray photons into visible photons is called fluorescence, the intensifying screen is one form of fluorescent screens. In fluoroscopy, the fluorescent screens are usually optimized for human visualization.

### 1.  Intensifying Screens

An intensifying screen is basically a layer of phosphor with a thickness range of 0.05 mm to 0.3 mm that emits light photons when struck by X-ray photons, with some supporting material. A detailed diagram depicting various components of an intensifying screen is shown in Fig. 16. The substrate material of approximately a few tenths of a millimeter thick should be relatively transparent to X-ray,

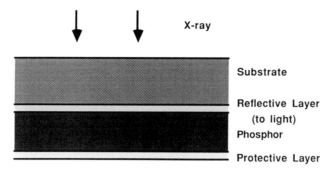

**Figure 16** Components of intensifying screen.

and the protective layer is light transparent. The light-reflective layer of 0.025 mm thick prevents the light photons from going in the backward direction. Since the X-ray photons are high-energy photons and light photons are low-energy photons, a small number of X-ray photons can generate a large number of light photons.

A number of phosphors have been used for intensifying screens. The most popular phosphor for intensifying screens has been calcium tungstate ($CaWO_4$) and terbium-activated rare-earth oxysulfide. $CaWO_4$, which has a K edge of 50.2 keV, emits light in the blue region of the visible light spectrum with a wavelength range of 350 to 580 nm and a peak wavelength of 430 nm, which is not optimal for visualization by the eye but very desirable for X-ray film exposure. The efficiency of $CaWO_4$ is relatively low, about 5%. The newer gadolinium ($Gd_2O_2S$), lanthanum (LaOBr), or yttrium oxysulfide phosphor ($Y_2O_2S:Tb$) can achieve a better efficiency of 15, 12, or 18%. The rare-earth phosphors are more absorbent in midrange energy between 50 and 60 keV than $CaWO_4$, and $Gd_2O_2S$ has a K edge at 50.2 keV. $CaWO_4$ screens have been largely replaced by rare-earth screens (Arnold, 1979). The rare-earth screens emit narrow bands of light in the green or blue region of the light spectrum which necessitates the use of a photographic film sensitive to light in these regions.

The speed of an intensifying screen for a given phosphor is determined primarily by its thickness, phosphor grain size, and the ability of the light photons to escape from the screen (Thomas, 1988). The thicker the phosphor, the higher the speed. This is because a thicker layer will absorb more X-ray photons and, thus, emit more light photons, which darken the film in a shorter time. However, the increase in thickness also results in a degradation in image quality, such as sharpness and contrast, due to more scattering or diffusion of the light photons. The effect of thickness on resolution can become clear by considering Fig. 17. A phosphor crystal closer to the film will blacken a smaller area of the film, whereas a crystal farther from the film will blacken a larger area. Precise definitions for image quality descriptors such as sharpness, contrast, and resolution can be found in Section V of this chapter. A light-absorbing dye may sometimes be added to the phosphor to improve the resolution by preferentially absorbing the light photons emitted laterally.

## 2. Image Intensifiers

The image produced on the image-intensifying screen is typically very weak and can be visualized only when the room is darkened. To brighten the image, a device called the image intensifier can be used. A typical image-intensifier tube is

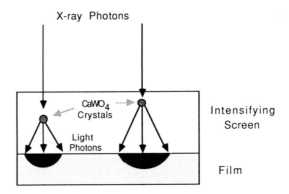

**Figure 17** Increase in screen thickness, although it increases screen speed, degrades image quality.

shown in Fig. 18. It is a vacuum tube with the following components: (1) input phosphor and photocathode, (2) focusing plates, (3) an anode, and (4) output phosphor. The X-ray photons that have propagated through the patient will be absorbed by the fluorescent screen of 15 to 35 cm diameter with almost spontaneous emission of light photons. The light photons strike the photocathode kept at ground potential, causing it to emit electrons in a number proportional to the brightness of the screen. The photocathode is usually made of a photoemissive metal such as antimony and cesium compounds. The electron beam will be accelerated and focused onto the output fluorescent screen by the anode and the focusing plates. The anode has a positive potential about 25 kV. The output fluorescent screen of 1.5 to 2.5 cm diameter, when struck by these high-energy electrons, produces an image much brighter than the image generated by a conventional

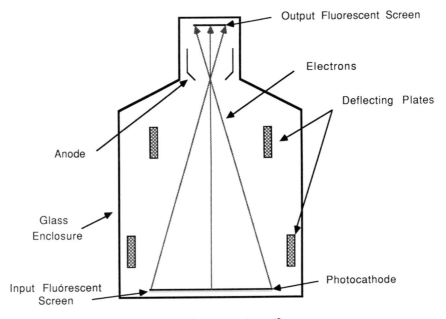

**Figure 18** Physical construction of an image intensifier.

fluoroscope with similar image quality because (1) the electrons gain energy during acceleration to the anode; and (2) the output screen is much smaller than the input screen. In the modern image intensifier, cesium iodide, which is more efficient and thinner than zinc cadmium sulfide, is used as the input phosphor and zinc cadmium sulfide is used as the output phosphor.

## F.  X-ray Detectors

Both films and radiation detectors have been used as X-ray receptors.

## 1.  X-ray Film

The X-ray film is a photographic film consisting of a transparent plastic substrate made of acetate or polyester usually coated on both sides with a light-sensitive emulsion. The reason for coating both sides will become clear later. The most important components of the emulsion are silver halide crystals with grain sizes from 0.1 to 1 $\mu$m, mostly silver bromide, and gelatin. The silver bromide crystal, upon receiving a light photon, yields a free electron that can combine with a silver ion to form a silver atom. Silver iodide 1–10% is used as a sensitizer. The silver atom is black. As a result, exposure of the film to light causes it to darken. The blackening process of the film is directly related to the energy of the incident light, which is the product of intensity of the light and time of exposure.

**Optical Density**   The photographic density or optical density used to measure the film blackness is defined by the following equation:

$$D = \mathrm{Log}_{10}(I_i/I_t) \tag{1-7}$$

where $D$ denotes the optical density, $I_i$ and $I_t$ are incident and transmitted light intensities, as illustrated in Fig. 19. This measure actually reflects the light opacity of the medium or the blackening of the film. Higher density means a darker film or less light transmission. In X-ray, a film with a density of 2 appears black when viewed on a standard light box, while a density between 0.25 to 0.3 appears to be white or transparent.

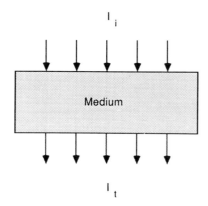

**Figure 19**   Presence of a medium causes attenuation of incident light intensity $I_i$. The ratio of transmitted light to incident light is used to measure opacity of medium.

An unexposed X-ray film after being processed shows a density usually less than 0.22, called the base plus fog density owing to (1) the base material on which the film emulsion is coated; and (2) silver halides developed without exposure, called film fog. One of the reasons for expressing the density in terms of a logarithmic scale is that the physiological response of the eye to the light intensity is logarithmic.

**Characteristic Curve**   The relationship between density and film exposure is called the characteristic curve, or the H and D curve, of the film. One such curve of an X-ray film with intensifying screen is shown in Fig. 20. It can easily be seen that even at zero exposure, the density is not zero because of the base material and film fog. In the mid-portion of the curve, which is the most important part of the curve, the density is approximately linearly proportional to the log exposure.

**Film Gamma**   Film gamma is defined as the maximum slope of the characteristic curve, or

$$\gamma = \frac{D_2 - D_1}{\log_{10} E_2 - \log_{10} E_1} \tag{1-8}$$

This is illustrated in Fig. 21(a). However, the average slope in the entire region of useful radiographic density (0.5 to 2.5) is more useful. Note here that a slope greater than 1 means the film amplifies the information carried by the light photons.

**Speed**   The speed or sensitivity of an X-ray film–screen combination is conventionally defined as the reciprocal of the X-ray exposure in roentgens required to

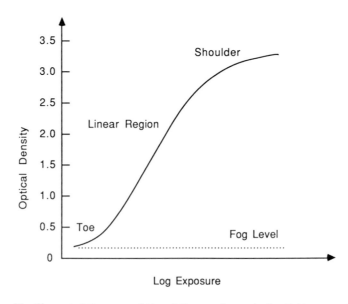

**Figure 20**   Characteristic curve of H and D curve (named after F. Hurter and V. C. Driffield of England) of photographic film.

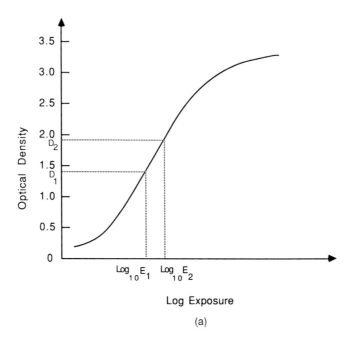

(a)

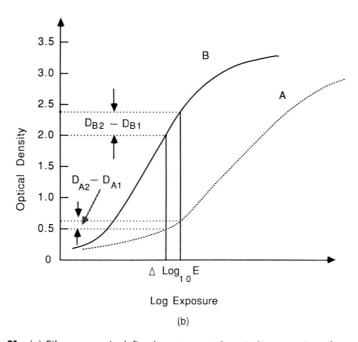

(b)

**Figure 21**  (a) Film gamma is defined as steepest slope in linear region of characteristic curve. (b) Film of larger latitude has lower film contrast.

produce a density of 1.0 above the base plus fog density of the film, or

$$S = 1/E$$

where $S$ is the speed and $E$ is the exposure in roentgens.

Its physical significance becomes evident by expressing $E$ in terms of the product of intensity and exposure time. If intensity is kept constant, a faster film, or a film with higher speed would require shorter exposure time to achieve an increase in optical density of 1. The speed of a film is dependent on the film development time, development temperature, silver grain size, and content (Thomas, 1988). An increase in development temperature is known to increase not only speed but also fog density.

**Film Latitude**   Film latitude is defined as the range of log exposure that produces acceptable optical density for diagnostic purpose (usually between 0.5 and 2.5). Large latitude results in low film contrast, defined as the difference in optical density between two adjacent regions. This is illustrated in Fig. 21(b), where film A has a larger latitude than film B but film contrast for A, or ($D_{A2} - D_{A1}$), is smaller than that for film B ($D_{B2} - D_{B1}$), for same $\Delta \log E$.

**Double-Emulsion Film**   The film used in daily radiological practice is coated on both sides of the film substrate. The reason is twofold: (1) It prevents bending of the film, because the film will bend if coated only on one side when the emulsion becomes dry and shrinks; and (2) it doubles film contrast. The configuration of a double-emulsion film cassette is shown in Fig. 22. The thicknesses for the emulsion and base are approximately 0.01 mm and 0.15 mm.

## 2.   Radiation Detectors

Two types of radiation detectors are currently used for X-ray detection: scintillation detectors and ionization chamber detectors. Figure 23 shows a scintillation detector, which consists of a scintillation crystal (e.g., sodium iodide with traces of thallium) coupled to a photomultiplier tube (see also Section II-C, Chapter 3). Scintillation crystals like sodium iodide emit light photons in proportion to the absorbed X-ray photon energy. The scintillation crystal surface is coated with a reflective material to collect the light photons.

The construction of a photomultiplier tube can also be seen in Fig. 23. It consists of a photocathode, an anode, and several intermediate electrodes (only two are shown in the figure) called dynodes. The photocathode is coated with a photoemissive material that emits electrons when stricken by light photons in proportion to the intensity of the light. Since the photocathode is maintained at the ground potential, which is much lower than the potential of the dynode closest to the scintillation crystal denoted as $V_1$, the emitted electrons or photoelectrons will be accelerated toward the dynode, which is covered by a material that emits secondary electrons when stricken by an electron. The potentials of the other dynodes are $V_2, \ldots, V_n$. In this way, the number of electrons is multiplied when they are propagating down the tube. Therefore, the output current is proportional to the number of light photons. The efficiency of this type of device is above 85%.

The second type of detector is the ionization chamber shown in Fig. 24. It consists of a chamber filled with a gas, usually xenon. Gas molecules in the the chamber are ionized by X-ray photons. The ions are then attracted to the elec-

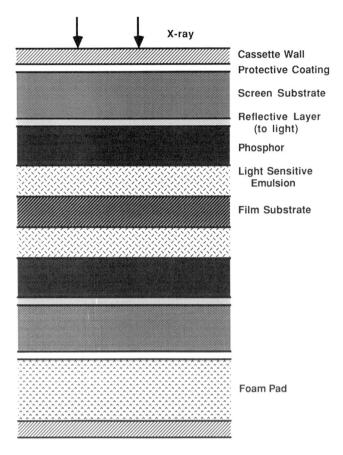

**Figure 22**  Components of double-emulsion film.

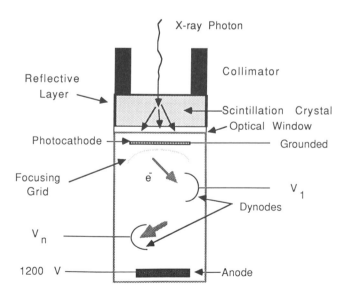

**Figure 23**  Physical construction of a photomultiplier tube.

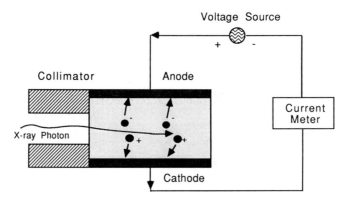

**Figure 24**   Physical construction of a radiation detector: ionization chamber.

trodes by a voltage difference between the electrodes, which is adjusted so as to produce a current accurately representing the energy of the X-ray photons absorbed. Because of the low density of the gases, some of the X-rays may travel through the chamber undetected. That is why xenon, which is the heaviest of the inert gases, is preferred. This type of device is relatively inefficient but cheap.

# III.   X-ray Diagnostic Methods

X-ray has been used in clinical diagnosis for many years. Therefore, there exist a variety of approaches, meeting specific requirements in different medical disciplines. In the following each approach will be described.

## A.   Conventional X-ray Radiography

Although there are now newer and more powerful diagnostic tools, X-ray radiography remains the most commonly used clinical procedure today. The reasons are that the radiologistis are familiar with the procedure, it is fully automated so that little training is required to operate the machine, its performance is superior to other modalities in a number of situations, and the image resolution is good. Figure 25 shows the basic components of a radiographic system, which were described in preceding sections. The rays produced by the generator are filtered to remove undesired energy and restricted to a certain cross section sufficient to illuminate the organ of interest (e.g., lung, head, etc.), for the purpose of minimizing patient exposure. Immediately behind the patient are positioned the grid for removing the scattered radiation and the screen–film combination for recording the image, which is a map of the transmitted X-ray intensity. Since X-ray attenuation in a tissue is proportional to electron density or mass density of the tissue, the X-ray intensity that has traversed a region of lower density (for instance, object O in Fig. 25) will be greater than that which has traversed a region of higher density. As a result, the area on the film behind the lower-density region is darker than the area behind the region of higher density. In other words, the gray level of the image is inversely proportional to the attenuation of the tissue in the ray path. It should be noted here that Eq. (1-4) is only valid for a homogeneous

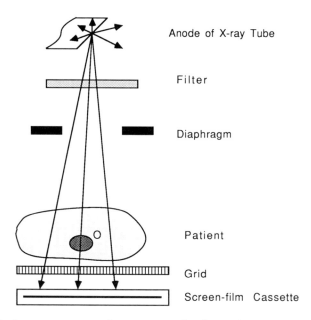

Anode of X-ray Tube

Filter

Diaphragm

Patient

Grid

Screen-film  Cassette

**Figure 25**  Basic components of a conventional radiographic system.

medium with constant attenuation coefficient $\beta$. For a heterogeneous medium like tissue, this equation no longer applies. This issue will come up when X-ray computerized tomography is discussed.

## 1.  Penumbra or Geometric Unsharpness

Since the focal spot of an X-ray generator is of finite size, a blurring effect on the image of an object will result, as illustrated in Fig. 26. A point object $P$ ideally should appear as a point on the film. However, because of the focal spot size $f$ the image is smeared or blurred. The width of this blurred image, $d$, is defined as the geometric penumbra or geometric unsharpness. $d$ is related to the focal spot size $f$ by the following equation:

$$d = ft/(S - t)$$

Therefore, to reduce the geometric unsharpness, $f$ and $t$ should be made small and $S$ should be large. It is plausible then to reduce image blurring due to the penumbra or geometric unsharpness: For a chest radiograph, the patient is asked to lean his or her chest as tightly as possible against the film cassette while the X-ray generator is located as far away as possibe from the patient.

## 2.  Field Size

As the X-ray beam leaves the focal spot, it diverges. If the beam restrictor size is $d_0$ as shown in Fig. 27, the field size $d_1$ at a distance $S_1$ from the focal spot from the geometrical relationship is given by

$$d_1 = d_0(S_1/S_0)$$

and the relationship between the size $d_2$ at a distance $S_2$ from the focal spot and $d_1$ is given by

$$d_2 = d_1(S_2/S_1)$$

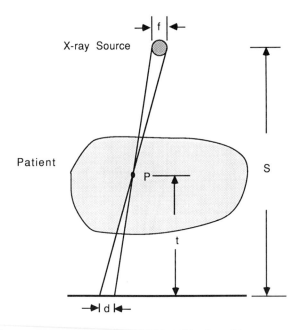

**Figure 26**   Finite aperture of X-ray source causes blurring of image.

Here it should be realized that the focal spot is finite and consequently there will
be penumbra associated with the boundaries. The above relationships hold if the
film is positioned at a large distance from the source.

## 3.   Film Magnification

Suppose that an object of interest, O, is located in the body of a patient and that
the object is located at a distance $t$ from the film (Fig. 28). It can easily be shown
that the size of the object, $L_0$, is magnified by a factor of

$$r_m = L_1/L_0 = S_f/(S_f - t)$$

where $L_1$ is the size of the object on the film and $S_f$ is the distance between the

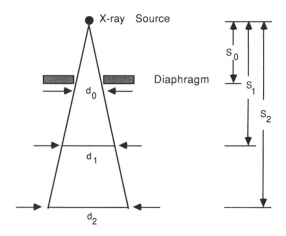

**Figure 27**   X-ray beam field size is proportional to distance from source.

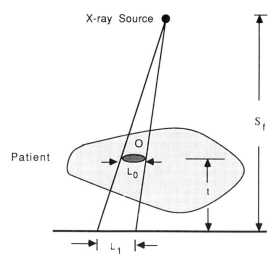

**Figure 28** X-ray image of object is magnified by ratio of $S_f/(S_f - t)$.

focal spot and the film. From this equation it becomes apparent that if multiple objects are imaged, the sizes of these objects on the film are distorted proportionally to the distances between these objects and the film. This distortion can be minimized if $S_f$ is $\gg t$.

X-ray is inferior to ultrasound or magnetic resonance imaging in differentiating soft tissues, but it is the method of choice for imaging bone and lung diseases because the density of these tissues is significantly different from soft tissues.

## B.  Fluoroscopy

The X-ray images can be recorded on a film for examination as was discussed or can be visualized directly on a flourescent screen if motion of the object of interest, such as a contrast medium like barium sulfate in the digestive tract, must be studied. A conventional fluoroscope is shown in Fig. 29, except that the image intensifier is replaced with a fluorescent screen assembly. The screen assembly consists of a grid, a slot for inserting the film if necessary, the fluorescent screen, and a lead glass layer to absorb radiation passing through the screen.

The problem with the conventional fluoroscope is that the image produced is very weak. So the room has to be darkened for examining the image. Even so, the quality of the fluoroscopic image is very poor compared to an average radiograph on film. Therefore, an image intensifier that amplifies the light signal is always used in the modern version of a fluoroscope (Fig. 29).

In a typical fluoroscopic procedure for examining the digestive tract, a contrast medium like barium sulfate that is nontoxic is either taken orally by the patient or by enema depending on which part of the GI tract is being examined. Common GI tract problems (e.g., ulcers, tumors, or obstructions) can be diagnosed by fluoroscopy. Figure 30 shows a colon radiograph where colon containing the contrast medium appears much darker than the surrounding tissues.

In this procedure the patient is being continuously exposed to X-ray radiation. The radiation dose received by the patient can be very high. It is therefore important to take this into consideration during an examination.

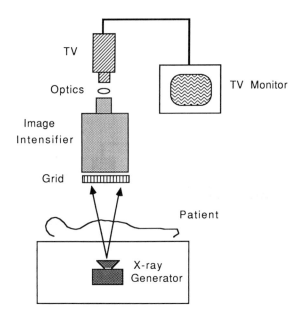

**Figure 29**   Basic components of fluoroscopy.

## C.  Angiography

Angiography is a procedure involving radiographic visualization of blood vessels by injecting a nontoxic radiopaque substance such as water-soluble organic compounds of iodine into the blood stream. It is useful for demonstrating vessel constriction and vascular tumors. A bolus of a contrast medium is injected into the artery or vein, usually through catheterization. The medium is then rapidly diluted in the blood circulation. A series of images are taken immediately after the injection. Figure 31 shows the cerebral angiographs taken at intervals of one second following the injection of the bolus. Injection of a foreign material into the body carries an inherent risk to the patient, that is, the patient may be allergic to the contrast agent. This is particularly true for agents like sodium iodide which is ionic. The fatality rate associated with angiography while using ionic contrast agents has been estimated to be in the order of one percent. The incidence of reaction may be reduced by using non-ionic contrast agents (Bettmann and Morris, 1986; Zelch, 1989). A major drawback with non-ionic agents is their exorbitant price.

Although many clinical applications of angiography have been replaced largely by noninvasive imaging techniques such as ultrasound and magnetic resonance imaging, angiography remains the "gold standard" for diagnosing a number of clinial disorders: Coronary stenosis, pancreatic disease, abdominal aortic aneurysm, and venous thrombosis are just a few examples (Keller and Routh, 1990).

## D.  Mammography and Xeroradiography

X-ray mammography is the radiographic examination of the breast performed with or without the injection of contrast medium. The requirements for mammography are different from ordinary X-ray examination for the following reasons:

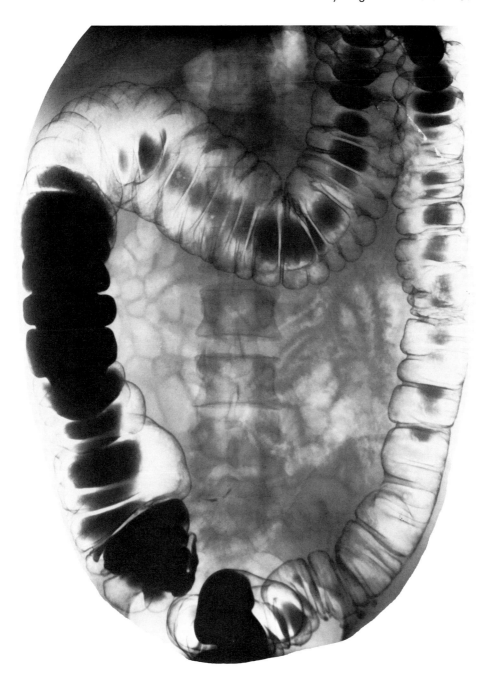

**Figure 30**  X-ray radiogram of colon of patient following air–barium double-contrast enema in which barium and air are used to coat and extend the GI tract. (Courtesy of John Cardella and Gary Thieme, Milton S. Hershey Medical Center, Penn State University).

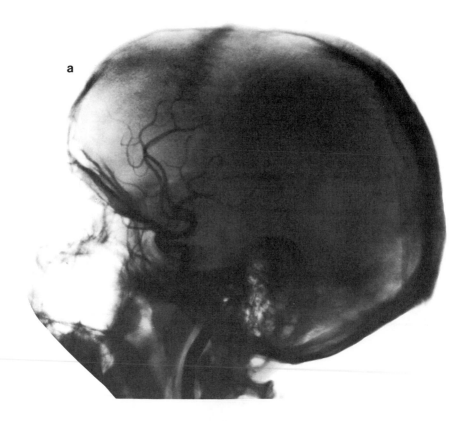

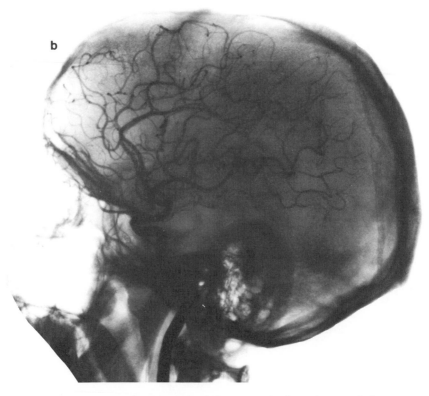

**Figure 31** Angiograms of cerebral circulation or cerebral arteriograms in 1-sec intervals following injection of contrast medium. (Courtesy of John Cardella and Gary Thieme, Milton S. Hershey Medical Center, Penn State University).

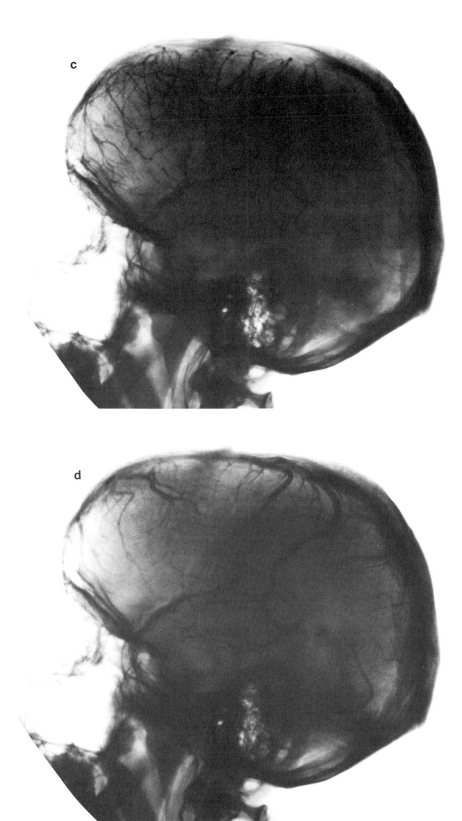

c

d

**Figure 31** (*continued*)

(1) Low-energy X-rays in the order of 20 keV should be used because the breast is composed of soft tissues. As can be seen in Fig. 6, the difference between attenuation coefficients of soft tissues is most pronounced at low X-ray energy levels. As a result modern mammography uses molybdenum (Mo) with an atomic number of 42 as anode material, which yields characteristic radiation peaking at 17.4 and 19.6 keV. (2) Mammography needs spatial resolution better than 0.1 mm to be able to visualize microcalcification and trabeculae. Thus, single-emulsion film that yields better resolution is preferred (e.g., Kodak Min-R/om-1™). (3) Exposure time should be short to avoid artifact due to patient motion. State-of-the-art film/screen mammography can achieve this resolution with 1/8 to 1/5 of the radiation dosage absorbed by the patient from Xeroradiography, which is discussed in the next section.

Xeroradiography is an X-ray technique developed by the Xerox Corporation that uses an X-ray energy level between 35 and 45 keV and an electrostatic technique to record the X-ray image instead of film. In this technique, a selenium-coated plate, which is positively charged, is exposed to X-rays coming through the patient. The X-ray photons cause the selenium to release electrons, neutralizing part of the positive charge. The areas of the plate under thick body parts will retain most of the positive charge, whereas on the areas of the plate under thin parts of the body, most of the charge will be neutralized. The plate is then sprayed with negatively charged blue powder. The amount of powder retained by a certain area of the plate will be an indicator of the X-ray intensity over that area. The powder pattern is transferred by heat to a sheet of paper coated with plastic for viewing and storage, as shown in Fig. 32.

During the spraying process, the powder near an area that has little remaining charge is attracted to the edge of the nearest area with more remaining charge, producing a well defined image of that edge. The edge enhancement effect is the main reason that the Xeroradiograph shows detail in thick body parts better than a conventional X-ray. However, Xeroradiography is less sensitive than conventional X-ray. To obtain a Xeroradiograph of similar quality, patient dose may have to be increased. In newer models in which a liquid toner is used the patient dose is cut to half (less than 0.3 rad for a 5 cm breast in a two-view study). A Xeroradiogram of a breast is shown in Fig. 33.

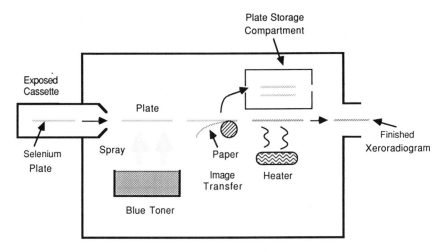

**Figure 32** Physical construction of Xeroradiographic system.

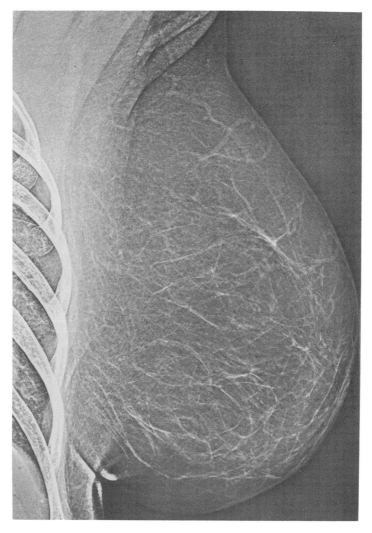

**Figure 33** Xeroradiogram of breast where ribs and breast vasculature are seen. (Courtesy of Xerox Medical Systems, Monrovia, California).

Both screen-film mammography and Xeromammography require compression of the breast for several reasons. Among the most important are: (1) maintaining a more uniform X-ray radiation on the breast; and (2) diminishing motion artifact and reducing radiation dosage since the breast tissue that needs to be penetrated is thinner (Bassett and Gold, 1987). Although significant progress has been made in X-ray mammography in increasing resolution and reducing patient exposure, recent reports show that 8 to 15% of breast masses (some of which are palpable, particularly in radio-dense breasts) may be missed (Kopans, 1989). Because of potential health hazards associated with ionizing radiation, other noninvasive imaging approaches for diagnosis of occult, early breast cancer have been sought and studied. The most notable are ultrasound, magnetic resonance imaging, light scanning, and thermography. Ultrasound mammography has been shown to be a valuable adjunct to X-ray mammography since it is superior in fatty breasts and is capable of differentiating solid tumors from cysts (Schoenberger, 1988).

## E.   Image Subtraction

Image subtraction is a procedure that is used to suppress background information on an image as first described by des Plantes in 1934 (des Plantes, 1934). It is achieved either digitally or by photographic process on the X-ray film. It is particularly useful in angiography because the image of the blood vessels under scrutiny is often masked by images of bone or soft tissue surrounding the vessels. This procedure is graphically illustrated in Fig. 34. Images in (a) and (b) all containing a large amount of information differ only in that (a) contains one more square pattern than (b). Just by examining the two images directly, it is very difficult to see the difference. A similar situation occurs in the angiograms illustrated in Fig. 31.

In image subtraction, a reversed image of (b) is produced. Then (a) and reversed (b) are superimposed and a final film in which the square pattern can be clearly seen is prepared as in (c). For temporal subtraction angiography, an initial image of the head is taken called scout film before a bolus of contrast medium is injected. The reverse of the scout film is called the mask. A second film is taken subsequently following bolus injection of a contrast medium. The superimposed image of the second image and the mask shows only the vascular structures so that the area of concern may become more visible (Fig. 35).

A major pitfall of the temporal subtraction technique is that the patient's head has to be kept motionless between exposures. This is almost impossible because of the sensation induced by the injection of the contrast medium. A compounding problem is the nonlinear characteristics of the photographic film. Therefore, perfect subtraction is rarely achieved in practice. Subtraction can also be done digitally and is termed digital subtraction angiography (DSA) (Kruger and Riederer, 1984). It is discussed in Section IV-A of this chapter.

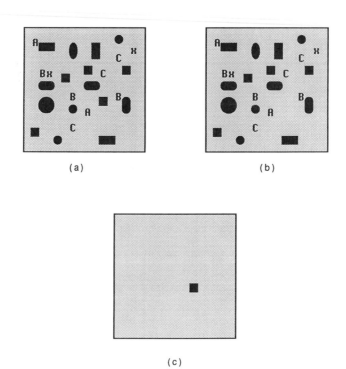

(a)                                    (b)

(c)

**Figure 34**   Schematic diagrams illustrating concept of image subtraction.

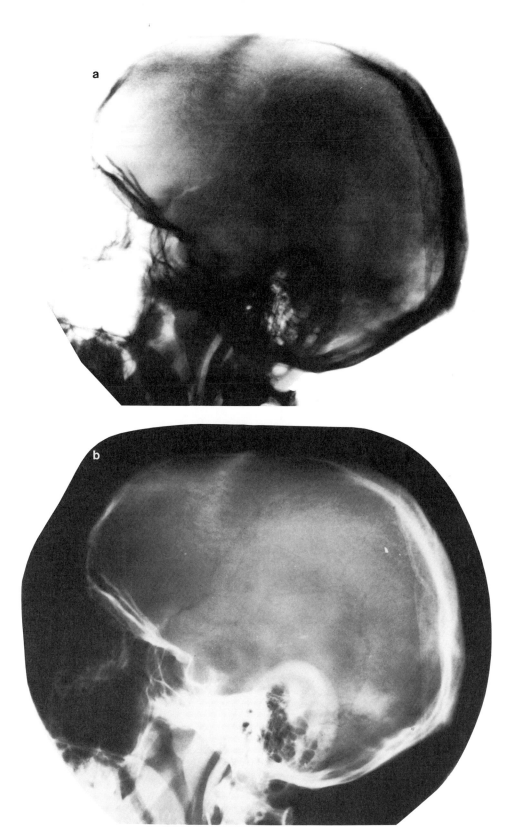

**Figure 35** Process of image subtraction: Obtain scout film of cerebral arteriogram (a) and mask (b), inject bolus of contrast medium and obtain arteriogram (c), and finally superimpose (c) on (b) to obtain subtracted image (d). (Courtesy of John Cardella and Gary Thieme, Milton S. Hershey Medical Center, Penn State University).

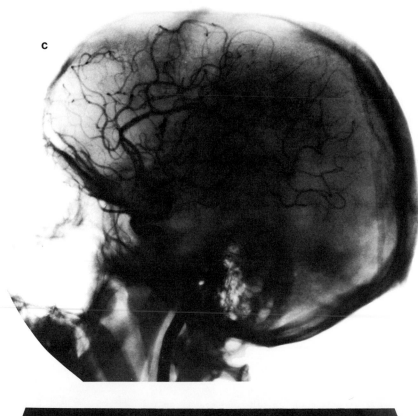

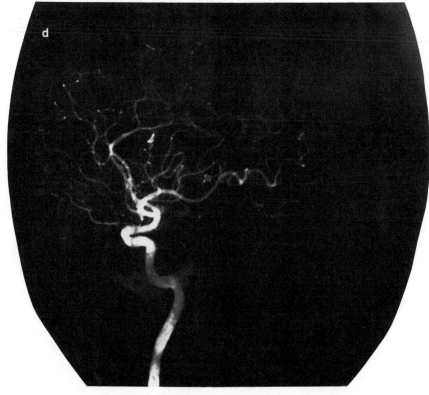

**Figure 35**   (*continued*)

## F.  Conventional Tomography

Conventional X-ray methods have a number of limitations. First, conventional X-ray is a two-dimensional projection of a three-dimensional structure. Many planes are superimposed onto one. Not only is the depth information lost, but more important, the confusion of overlapping planes makes the detection of subtle abnormalities very difficult if not impossible. Second, conventional X-ray cannot differentiate between soft tissues. As has been discussed, a radiograph is valuable for imaging tissues whose densities differ significantly from soft tissues. Imaging of certain soft tissues can be accomplished only through the use of radiopaque dyes. Third, conventional X-ray cannot be used to measure in a quantitative way the densities of the various tissues being interrogated.

A partial solution to these problems is to use the tomographical imaging technique, or tomography (the Greek word *tomo* means cut). X-ray tomography is a special X-ray technique that blurs out undesirable images of superimposed structures in order to accentuate the images of principal interest. We emphasize that this technique does not improve the resolution of the image but merely blurs the undesirable parts. This procedure was first described by Bocage of France (Laughlin, 1988).

The tomographic principle is depicted in Fig. 36. The essential parts are an X-ray tube, an X-ray film, and a metal structure that rotates about a pivot point or a fulcrum. The film moves in synchronization with the motion of the X-ray tube, but in the opposite direction. The plane containing the pivot point and parallel to the X-ray film is the plane of interest (fulcrum plane), which is the only plane in sharp focus. All points above and below this plane are blurred, as seen in Fig. 36. Suppose that there are two objects in the patient, represented by a square and a circle. The circle is on the fulcrum or focal plane. As the X-ray tube is translated continuously to the left, the X-ray film is moved to the opposite direction. The speed of the film is so adjusted that the image of the circle falls on the same

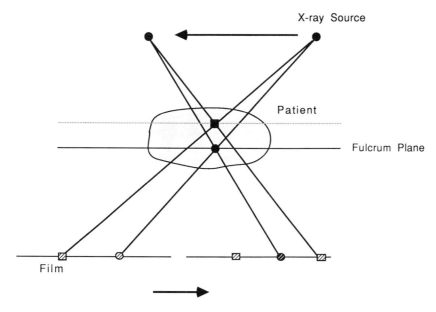

**Figure 36**  Conventional tomography.

spot as the X-ray tube is moved. So only the image of the circle remains in sharp focus. The image of the square is blurred out because it moves at a faster speed than the circle and as a result is not in synchronization with the movement of the X-ray tube. This arrangement is termed "linear tomography" in which the source and detector move linearly in opposite directions. The degree of blurring depends on the distance that the source and detector have travelled, the distance between the object to be blurred out and the focal plane, and the orientation of the structure of interest. Since images of structures parallel to the direction of tube movement will not be blurred out, a more complex circular or elliptical motion of the tube is employed (Curry *et al,* 1990). As expected, the image quality of the tomogram produced is relatively poor since the blurred images of undesirable structures add to the background noise.

## G. Computed Tomography

Conventional tomographic techniques solve the first problem of conventional X-rays; that is, structures on a certain plane can be imaged. However, these techniques still cannot circumvent the second and the third problem. Computed tomography, on the other hand, offers so much greater sensitivity that soft tissue can be differentiated, and it also provides quantitative informaiton about the attenuation properties of tissues through which the X-ray beam has traversed.

The idea of computed tomography (CT) is really not very new. The mathematical basis for this approach existed as early as in 1917 (Radon, 1917), but the practical application did not start until the early 1960s. Allan Cormack, at Tufts University, was one of the people during that period popularizing the idea (Cormack, 1963). However, the first practical scanner was not built until 1972 by Hounsfield, at EMI in England (Hounsfield, 1973). Cormack and Hounsfield were awarded the Nobel Prize for their contribution to this field in 1979.

The basic physical principle involved in CT is that the structures on a 2-D object can be reconstructed from multiple projections of the slice. Before illustrating this by a simple example, let us define the projection function. Suppose that a 2-D slice of an object, whose attenuation coefficient at a point $r(x, y)$ is denoted by $\beta(r) = \beta(x, y)$, is illuminated by a uniform beam of X-ray in a direction of $\phi$ relative to a reference coordinate $(x, y)$, as shown in Fig. 37. Coordinate system $(x', y')$ makes an angle $\phi$ with respect to the reference coordinate system $(x, y)$. If the object is homogeneous so that the attenuation coefficient is a constant independent of the spatial position, Eq. (1-4) can be used.

### 1. Projection Function

For inhomogeneous materials such as biological tissues the linear attenuation coefficient $\beta$ is no longer a constant. Now if the incident beam is in the $y'$ direction and we consider that the object consists of many small squares, each with a dimension, say, $\Delta y'$ along the $y'$-axis with the $x'$ coordinate being $x'_1$, then we have along $y'$, assuming there are $n$ squares along $y'$ at $x'_1$,

$$I = I_0 e^{-[\beta(x'_1, y')\Delta y' + \beta(x'_1, y' + \Delta y')\Delta y' + \beta(x'_1, y' + 2\Delta y')\Delta y' + \cdots]}$$

or

$$I = I_0 e^{-\sum_{i=1}^{n} \beta(x'_1, y' + i\Delta y')\Delta y'} \tag{1-9}$$

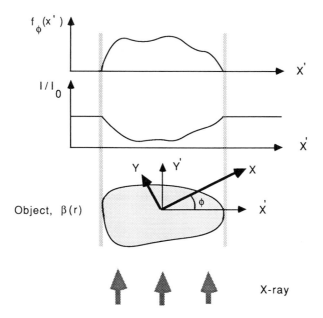

**Figure 37**  Projection function of object at angle $\phi$ relative to reference coordinate system (X, Y).

We can easily express the summation in terms of an integral, that is,

$$I = I_0 e^{-\int_{\text{Source}}^{\text{Detector}} \beta(x_1', y')dy'} \qquad \text{at } x' = x_1'$$

Since $x_1'$ is arbitrary, we have

$$I(x') = I_0 e^{-\int_{\text{Source}}^{\text{Detector}} \beta(x', y')dy'} \qquad (1\text{-}10)$$

Now

$$x' = x \cos \phi + y \sin \phi$$
$$y' = -x \sin \phi + y \cos \phi$$

Substituting these relations into Eq. (1-10), it can be seen that

$$I(x') = I_0 e^{\int_{\text{Source}}^{\text{Detector}} \beta(x, y)dy}$$

or

$$I/I_0 = \exp\left[ -\int_{\text{Source}}^{\text{Detector}} \beta(x, y)\, dy \right]$$

Taking negative logarithms of both sides of the equation, we have

$$f_\phi(x') = -\ln(I/I_0) = \int_{\text{Source}}^{\text{Detector}} \beta(x, y)\, dy \qquad (1\text{-}11)$$

where $f_\phi(x')$ is called the projection function, which is depicted in Fig. 37.

In practical situations, the body is surrounded by air, which has a relatively low attenuation coefficient, and Eq. (1-11) can be simply written as

$$f_\phi(x') = \int_{-\infty}^{\infty} \beta(x, y)\, dy \qquad (1\text{-}12)$$

Now suppose that the projection functions of an object as shown in Fig. 38 are known. The projection function can be measured by illuminating the object with X-ray and moving a detector, either a scintillation detector or an ionization chamber detector, along the $x'$ direction. The original object can be reconstructed by backprojecting the projection function as illustrated in Fig. 39 and appropriately adjusting the gray scale. Here this is simply used as an example to show how the image can be reconstructed from back projection. A more rigorous mathematical treatment will be given when reconstruction algorithms are discussed.

Alternatively, the CT concept can be better understood by considering the following approach. Here we are interested in determining the distribution of $\beta$ on a slice of an object as shown in Fig. 40. Let us say that the slice can be divided into four compartments, for the sake of simplicity. Each individual 2-D element is called a pixel. In reality, an X-ray beam has a finite width since it cannot be made infinitely thin. These elements are 3-D elements and should be called voxels. Let us assume an ideal case of an infinitely thin X-ray beam. Each pixel is assigned an attenuation coefficient. From Eq. (1-4), we have along direction $x$ and $y$ four equations from which the four unknowns, $\beta_{11}$, $\beta_{12}$, $\beta_{21}$, and $\beta_{22}$, can be readily solved.

$$I_1 = I_0 e^{-(\beta_{11}+\beta_{12})\Delta x}$$
$$I_2 = I_0 e^{-(\beta_{21}+\beta_{22})\Delta x}$$
$$I_3 = I_0 e^{-(\beta_{11}+\beta_{21})\Delta y}$$
$$I_4 = I_0 e^{-(\beta_{12}+\beta_{22})\Delta y}$$

where it may be assumed that $\Delta x \approx \Delta y$. This assumption is particularly valid when the number of pixels is large. It has to be noted that these four equations are not independent. More data are needed to find a correct solution. The solution of these simultaneous equations can be performed by a computer. Solution for a $512 \times 512$ matrix can be calculated by a microcomputer within a few seconds.

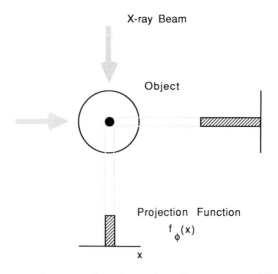

**Figure 38**   Projection functions of an object of simple geometry at different angles.

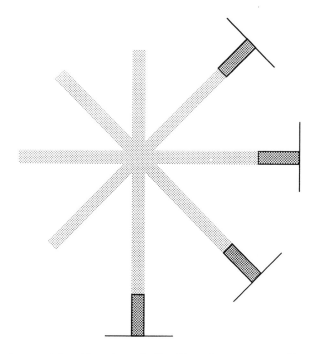

**Figure 39** Image of the object shown in Fig. 38 can be reconstructed by backprojection.

Two correction factors are generally incorporated into the CT program to compensate for the polychromatic nature of the X-ray beam and the geometrical problem because the scanning rays may traverse the pixels obliquely. Since the attenuation coefficient of a tissue is dependent on the energy of the X-ray photons, the lower-energy photons are absorbed and only higher-energy photons remain as the beam penetrates the body. This phenomenon is called "beam hardening" and has to be compensated. In the latter case, a weighing factor (WF) is used to correct for the differences in the propagation length, as depicted in Fig. 41.

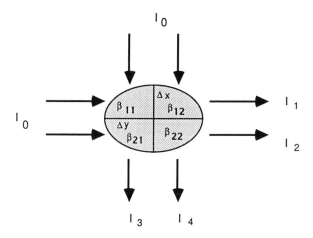

**Figure 40** Object divided into four pixels.

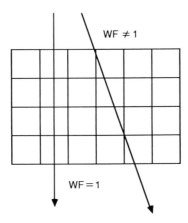

**Figure 41**   Weighing factor (WF) has to be applied to account for differences in propagation distance of X-ray beam in each pixel at different angles.

Three types of CT machines currently being used are shown in Fig. 42. Figure 42(b) shows a second generation unit. The gantry of the system translates across the body, each detector taking parallel sets of readings; at the same time it rotates around the body. It could have 30 detectors and require 18 seconds to obtain an image. A third generation machine is shown in Fig. 42(c). The sets of readings are taken in the form of a fan beam. It does not translate across the body but only rotates around it. This type of machine usually has 500 to 700 detectors but is much faster and can take a picture in as little as three seconds. (Examples of third generation scanners are GE 9800 and Siemens DR3.) The fourth generation system is shown in Fig. 42(d) (The Picker 1200 GK is an example.). The detectors are assembled in a fixed circle and only the X-ray tube sweeps around the body. The system has 500 to 1000 detectors and it can acquire a picture in as little as two seconds. However, it has only one moving part. The first-generation EMI machine with only one X-ray source and one detector shown in Fig. 42(a) required about five minutes to complete a scan, and now a scan can be completed within two seconds. The radiation, therefore, is substantially reduced.

A modern CT scanner is shown in Fig. 43. The X-ray generator and detectors are all housed in a gantry that has a circular opening where the patient is positioned. The image of a 2-sec, 10-mm slice is shown in an abdominal scan (Fig. 44) where gallbladder, spleen, kidney, and associated vasculature are clearly seen. Current CT machines typically have a voxel thickness of approximately 1 cm and are capable of resolving objects as small as 1 mm under ideal conditions.

Typically, X-ray tubes with 110–150 kVp are used in CT scanners. The X-ray beams are filtered and collimated to a narrow width in the order of 1 cm. Both ionization chamber detector and scintillation detector have been used in third generation scanners. Fourth generation scanners almost exclusively use scintillation detectors. It is imperative that all X-ray detectors have similar characteristics in CT scanners. Any mismatch in detector characteristics can result in severe image artifacts and distortion. It is for this reason that although fourth generation scanners, which in general have more detectors than third generation scanners,

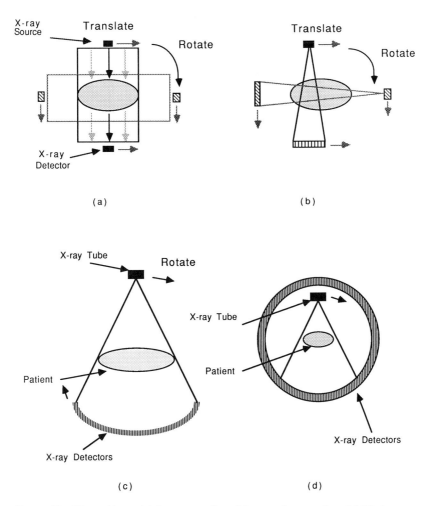

**Figure 42** CT machines: (a) first generation, (b) second generation, (c) third generation, and (d) fourth generation.

possess the advantage of one moving part, third generation scanners remain popular among the manufacturers. The gantry can be tilted within ±30° angle to obtain images at oblique angles. Data acquisition and image reconstruction of a CT scanner are all done digitally by one or more microprocessors which range from 16-bit DEC LSI to 32-bit Motorola 68000. Two hundred and fifty-six gray levels are commonly used for displaying the image of 512 × 512 to 1,024 × 1,024 pixels. The pixel size is in the order of 0.05 to 1.8 mm.

Image reconstruction for conventional CT scanners is performed off-line after all the data have been acquired. Typically the reconstruction time per scan is approximately between 2–20 sec. The amount of time needed to obtain multiple slices of image is considerable (3–30 scans/min). Fast CT scanners that use alternative designs to reduce image acquisition time, called cine-CT, for imaging moving structures like the heart, have been developed and will be discussed in Section IV of this chapter.

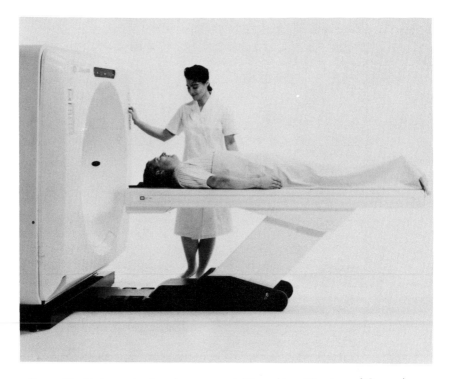

**Figure 43**   Photograph of modern compact CT machine. (Courtesy of General Electric Medical Systems, Milwaukee, Wisconsin).

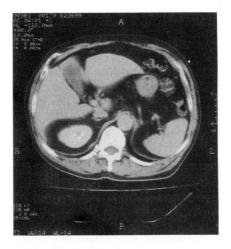

**Figure 44**   CT image of abdomen. (Courtesy of General Electric Medical Systems, Milwaukee, Wisconsin).

## 2. Algorithms for Image Reconstruction

Thousands of equations have to be solved to obtain a CT image. The best algorithms will be the ones that solve the equations as rapidly as possible. Four types of methods have been used for image reconstruction. They are (1) back projection, (2) iteration, (3) Fourier transform, and (4) filtered back projection.

**Back Projection Method**   The back projection method, or the summation method, was illustrated in the beginning of this section with a simple example. Here a mathematical treatment is given. If the back projection function of a projection function, $f_\phi(x', y')$ in the $y'$ direction is given by $g_\phi(x', y')$, which involves the smearing of $f_\phi(x', y')$ along the $y'$ axis, we have

$$g_\phi(x', y') = f_\phi(x') \qquad \text{for all } y'$$

In polar coordinates as shown in Fig. 45,

$$g_\phi(r, \theta) = f_\phi[r \cos(\theta - \phi)]$$

which follows from $x = r \cos \theta$, $y = r \sin \theta$, $x' = x \cos \phi + y \sin \phi = r \cos(\theta - \phi)$ and $y' = -x \sin \phi + y \cos \phi = r \sin(\theta - \phi)$.

Following back projection of $n$ projection functions as illustrated in Fig. 39, the discretely summed image $\Omega(r, \theta)$ can be represented by

$$\Omega(r, \theta) = \frac{1}{n} \sum_{i=1}^{n} g_{\phi i}(x', y') = \frac{1}{n} \sum_{i=1}^{n} f_{\phi i}[r \cos(\theta - \phi_i)] \qquad (1\text{-}13)$$

Since as many back projections as desired can be made, or $n$ can be extremely large, it is reasonable to approximate the summation in Eq. (1-13) by an integral.

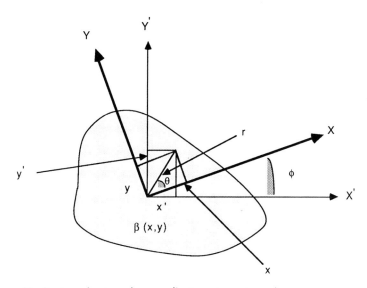

**Figure 45**   Rectangular-to-polar coordinate system conversion.

$$\Omega(r,\ \theta) = \frac{1}{\pi} \int_0^\pi g_\phi(x',\ y')\ d\phi$$

$$= \frac{1}{\pi} \int_0^\pi f_\phi[r\cos(\theta - \phi)]\ d\phi$$

Substituting Eq. (1-12) into this equation and noting that $y'$ is a function of $y$, we have

$$\Omega(r,\ \theta) = \frac{1}{\pi} \int_0^\pi d\phi \int_{-\infty}^\infty \beta(x',\ y')\ dy' \tag{1-14}$$

For a point object, $\beta(x',\ y') = \delta(x')\ \delta(y')$ where $\delta$ represents delta function. Equation (1-14) becomes

$$\Omega(r,\ \theta) = \mu(r,\ \theta) = \frac{1}{\pi} \int_0^\pi d\phi \int_{-\infty}^\infty \delta(x')\ \delta(y')\ dy'$$

$$= \frac{1}{\pi} \int_0^\pi \delta(x')\ d\phi = \frac{1}{\pi} \int_0^\pi \delta[r\cos(\theta - \phi)]\ d\phi$$

This integration can be carried out and is found to be (Barrett and Swindell, 1981)

$$\Omega(r,\ \theta) = \mu(r,\ \theta) = 1/(\pi r) \tag{1-15}$$

where $\mu(r,\ \theta)$ is the point response function.

Now from linear system theory, the output $o$ of a linear system with input $s$ is given by

$$o(u_1,\ u_2) = h(u_1,\ u_2)\ ** \ s(u_1,\ u_2) \tag{1-16}$$

where $u_1$ and $u_2$ are arbitrary variables, $h$ is the impulse response of the linear system, and the symbol ** denotes 2-D convolution which mathematically is represented by the following expression.

$$o(u_1,\ u_2) = \int_{-\infty}^\infty \int_{-\infty}^\infty s(\tau_1,\ \tau_2)h(u_1 - \tau_1,\ u_2 - \tau_2)\ d\tau_1\ d\tau_2$$

In the Fourier domain,

$$O(\omega_1,\ \omega_2) = H(\omega_1,\ \omega_2)S(\omega_1,\ \omega_2) \tag{1-17}$$

where $O$, $H$, and $S$ are the Fourier transforms of $o$, $h$, and $s$. The Fourier transform of a function, $o(u_1,\ u_2)$, is given by the mathematical expression:

$$O(\omega_1,\ \omega_2) = \int_{-\infty}^\infty \int_{-\infty}^\infty o(u_1,\ u_2)e^{j2\pi(u_1\omega_1 + u_2\omega_2)}\ du_1\ du_2$$

The function $H$ is also called the transfer function of the system. $\omega_1$ and $\omega_2$ are the corresponding variables of $u_1$ and $u_2$ in the Fourier domain.

If the back projection process is considered a linear system process, it can easily be seen that for an object with an attenuation coefficient distribution $\beta(x',\ y')$, the back projection function $\Omega(x',\ y')$ is related to the point response $\mu(x',\ y')$ by the following relationship:

$$\Omega(x',\ y') = \mu(x',\ y')\ ** \ \beta(x',\ y') \tag{1-18}$$

In the Fourier domain,

$$\Lambda(\rho, \xi) = M(\rho, \xi)\Delta(\rho, \xi) \qquad (1\text{-}19)$$

where $\Lambda$, $M$, and $\Delta$ are spatial Fourier transforms of $\Omega$, $\mu$, and $\beta$; $\rho$ and $\xi$ are spatial frequencies. Now if $R = (\rho^2 + \xi^2)^{1/2}$ and we know the Fourier transform of $1/r$ is $1/R$, we have

$$\Delta = \pi R \Lambda$$

In other words, $\beta$ can be estimated from the inverse Fourier transform of $\pi R \Lambda$ where $\Lambda$ is the Fourier transform of the back projection function $\Omega$ that can be obtained experimentally. This operation, $\Lambda$ multiplied by $R$, represents a spatial filtering process with the filter transfer function given in Fig. 46. However, in practical situations, the utilization of this type of filter is not feasible because high-frequency noises are amplified unproportionally, which can drown out the actual image information. Therefore, in practical applications, some forms of apodized filters, shown in the dotted line in Fig. 46, are used.

**Iteration Method**    Various types of iterative methods are available (Swenberg and Conklin, 1988). Basically they are very similar in principle. Essentially in these methods a set of initial values are assigned to all pixels. These values are then substituted into the set of simultaneous equations to see whether these equations are satisfied. If not, an algorithm is used to change these values, which are then tested again to determine whether the equations are satisfied. The process is repeated over and over again until a solution is found. This type of method is therefore quite computing intensive.

**2-D Fourier Transform Method**    The two-dimensional Fourier analysis technique needs to be elaborated a little bit more to be able to understand it. The one-dimensional Fourier transformation of the projection is

$$F_1[f_\phi(x')] = \int_{-\infty}^{\infty} f_\phi(x')e^{j2\pi\rho x'} \, dx'$$

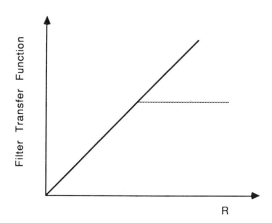

**Figure 46** Spatial filter transfer functions for backprojection where meshed line indicates apodized filter.

Substituting into this equation the definition of $f_\phi(x')$, we find

$$
\begin{aligned}
F_1[f_\phi(x')] &= \int_{-\infty}^{\infty} dx' \int_{-\infty}^{\infty} dy' \, \beta(x', y') e^{j2\pi\rho x'} \\
&= \left[ \int_{-\infty}^{\infty} dx' \int_{-\infty}^{\infty} dy' \, \beta(x', y') e^{j2\pi(\rho x' + \xi y')} \right]\Bigg|_{\xi=0} \\
&= F_2[\beta(x', y')]|_{\xi=0}
\end{aligned}
$$

where

$$
F_2[\beta(x', y')]|_{\xi=0}
$$

is simply the two-dimensional Fourier transform of $\beta(x', y')$ evaluated along the line $\xi = 0$ in a two-dimensional frequency space. This is the mathematical representation of the so-called central slice theorem. In brief, this expression says that a one-dimensional Fourier transform of a one-dimensional projection of a two-dimensional object is mathematically identical to one slice through the two-dimensional Fourier transform of the object itself. Thus, knowledge of all one-dimensional projections is sufficient to synthesize the two-dimensional transform of the object from which the image of the object is readily obtainable by an inverse of the two-dimensional transform.

In summary, to reconstruct the image with the 2-D Fourier transform method, (1) 1-D projections of the object along many directions are taken, (2) 1-D Fourier transforms of these 1-D projections are obtained, (3) all 1-D transforms are assembled to form a 2-D transform according to the angles of the projections relative to a reference coordinate (Fig. 47), and finally (4) the image is reconstructed from an inverse transform of the 2-D Fourier transform (Huang, 1987).

**Filtered Back Projection Method**   As discussed in the section on back projection method, the image of the object can be reconstructed from the filtered summed

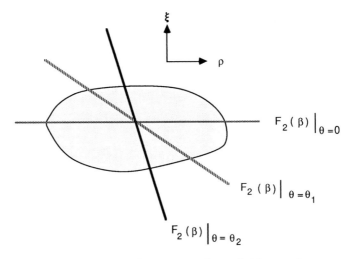

**Figure 47**   Two-dimensional spatial Fourier transform of object can be reconstructed from multiple Fourier transforms of 1-D projection functions obtained at different angles of object.

back projections. Since summation and filtering are all linear operations, they can be interchanged. The advantage of doing this is that the 2-D filtering process is reduced to 1-D operation, which saves considerable computation time. This is the primary reason that a majority of the CT equipment manufacturers today prefer this algorithm over others.

In essence, this approach requires that each projection be convolved with a 1-D filter function $q_1(x')$ before summation.

$$f_\phi^+(x') = \int_{-\infty}^{\infty} f_\phi(x_1) q_1(x' - x_1) \, dx_1 = f_\phi(x') * q_1(x')$$

where $f^+$ represents the convolved projection function and * denotes 1-D convolution. Therefore, proceed as before

$$\Omega^+(r, \theta) = \frac{1}{\pi} \int_0^\pi g_\phi^+[r \cos(\theta - \phi)] \, d\phi$$

where $g_\phi^+$ is the back projection of the filtered projection function as opposed to $g_\phi$ associated with straight back projection and $\Omega^+$ indicates the summed back projections of the filtered projection functions in contrast to $\Omega$. It can be shown (Barrett and Swindell, 1981) that the point response $\mu^+$ in this case is given by

$$\mu^+(r, \theta) = \frac{1}{\pi} \int_0^\pi q_1[r \cos(\theta - \phi)] \, d\phi$$

The filter function typically used is the Ramachandran–Lakshminarayanan filter or the Shepp–Logan filter, which is shown in Fig. 48.

## 3. CT Number

We have discussed how the data can be collected and the image can be reconstructed. The next topic should be how to display the image. The linear attenuation coefficients crunched out by the computer are usually given in terms of values relative to that of water, magnified to a larger integer by multiplying the normalized difference by a larger integer, called a CT number, or

$$\text{CT Number} = K \frac{\beta_p - \beta_w}{\beta_w}$$

where $K$ is an arbitrary integer (usually $K = 500$ or $1000$), $\beta_p$ is the linear attenuation coefficient of a certain pixel, and $\beta_w$ is the linear attenuation of water. Therefore, the CT number for water is zero. For $K = 1000$, the CT numbers of dense bone and air are respectively 1000 and $-1000$. All other human tissues have intermediate values, as shown in Table II. The accuracy of a modern CT scanner in terms of CT number is $\pm 2.0$ with $K = 1000$. CT numbers are given the unit Hounsfield when $K = 1000$. A gray scale or a color scale can then be used to represent the CT number or attenuation coefficient. As with digital radiography, which will be discussed later, data acquisition, processing, and storage in a CT scanner are all accomplished by a computer that allows the flexibility of windowing in gray scale display (Fig. 49). This flexibility is particularly useful when displaying soft tissues where the attenuation coefficient varies only in a limited range.

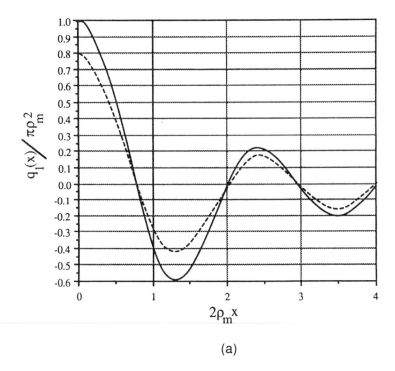

(a)

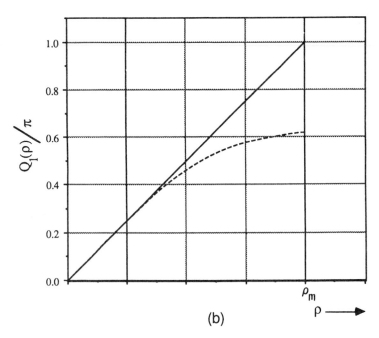

(b)

**Figure 48**   (a) Spatial filters for filtered backprojection. Solid and dashed lines represent respectively R–L (Ramachandran–Lakshminarayanan) filter and S–L (Shepp–Logan) filter. (b) Corresponding frequency responses; $\rho_m$ is the spatial cutoff frequency of these filters.

**Table II**
CT number and attenuation coefficients of
pertinent media and biological tissues
($K = 1000$, 60 keV photons)

| Medium | CT number | Linear attenuation coefficient, $cm^{-1}$ |
|---|---|---|
| Bone | 808 | 0.38 |
| Water | 0 | 0.21 |
| Striated muscle | −48 | 0.20 |
| Fat | −142 | 0.18 |
| Air | −1000 | 0 |

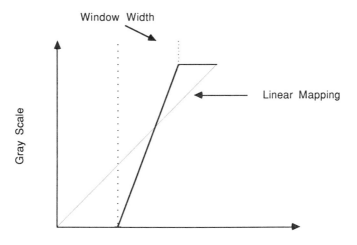

**Figure 49**  Windowing of gray-scale display.

## 4.  Image Artifacts

A number of problems can occur during a CT scan to result in image artifacts (Curry *et al.*, 1990). Patient motion during scanning can result in a blurring of the image since a finite amount of time is needed to acquire the image. Beam hardening causes inaccuracy in the estimation of attenuation coefficients in deeper pixels, particularly in pixels behind attenuating structures such as bone. Dark streaky shadows often appear behind bony structures in a CT image. An object of high density such as a bullet may appear as a star pattern of streaks because of the large difference in attenuation coefficient between the object and surrounding biological tissues. There is essentially no attenuation information for pixels behind this object since the X-ray energy is almost completely blocked by the object.

CT spatial resolution or its ability to resolve small objects is dependent upon many factors including the number of pixels, slice thickness, algorithms used, data sampled, and hardware. For a circular field of view of 20 cm diameter with $256 \times 256$ pixels, each pixel represents approximately $4.8 \times 10^{-3}$ cm$^2$. Spatial resolution may improve by increasing the number of pixels. However, a further increase in the number of pixels would not improve the resolution if the number of pixels exceeds the Nyquist rate or one half of the number of data sampled.

Computed tomography eliminates problems associated with conventional radiography. Namely, it gives a 2-D image of a slice; it is capable of differentiating soft tissues (e.g., a tumor from normal tissue), and it provides a quantitative assessment of the tissue in terms of attenuation coefficient. Therefore, CT has been used in diagnosis of a variety of clinical problems. The most important contribution of CT, however, is in the diagnosis of brain and spine disorders, such as tumor, hemorrhage, localization of foreign objects, etc. The clinical applications of CT in the abdomen, chest, and extremities are being challenged by magnetic resonance imaging.

# IV. Recent Developments

## A. Digital Radiography

The advent of digital computers and integrated circuits has made image processing and acquisition in digital format at a reasonable cost and within a short period of time a reality. In the late 1970s at the heels of the success achieved by CT, digitization of radiography seemed inevitable. Following more than a decade of research and development, it has been realized that we are still years away from completely replacing conventional radiography with digital radiography, because more components, which produce more noise, have to be introduced in the digitization process and the quality of a digitized image is not comparable to that of the conventional film at similar cost. Digitization of image has been achieved in two ways: (1) As in CT, a detector is used to scan the X-ray field, and (2) an area detector is used to capture the image one frame at a time (Brody, 1984, and Huang, 1987). The disadvantages of the former approach are that scanning time is long so that patient motion becomes a problem, and it is more expensive. The methods that have been used to accomplish the latter approach can be divided into two categories: computed radiography and digital fluoroscopy.

Conventional radiography uses a film/screen combination to capture the X-ray image, whereas computed radiography uses an imaging plate system that consists of an imaging plate, which is a photostimulable phosphor screen made of europium-activated barium fluorohalide compounds, and a scanning mechanism introduced in 1983 by Fuji (Huang, 1987). The gray scale of the developed film can be quantitated by scanning the film with a variety of means, such as a video camera, CCD (charge-coupled device) camera, drum scanner, and laser scanner. The output of the scanning device may then be digitized by an ADC (analog-to-digital converter) to create a digital image. This process is slow and time consuming. The imaging plate system on the other hand is fully automated under computer control. The imaging plate, which has a larger latitude than film, stores the X-ray image. The recorded image on the plate is extracted by scanning, usually by a He–Ne laser beam, to stimulate phosphor crystals at a quasi-stable state after absorbing the X-ray photons to emit light photons. The emitted light is collected by a lens and detected by a photomultiplier tube, whose output is then digitized. The imaging plate can be reused after erasing the stored image by flooding the plate with light. The major advantage of computed radiography is that the image information is available in digital format that allows easy implementation of image processing algorithms on the image such as spatial filtering, signal compression, and fast image storage and transmission. It has been described as one of the missing links in the full implementation of picture archiving and communication systems (PACS) needed for the total digitization of a radiology department. However, the cost of such devices has prevented it from wide acceptance. Alternatively the X-ray image may be captured directly by an image intensifier as in fluoroscopy, and the image on the output screen of the image intensifier is focused and scanned by a video camera. The video signal from the camera is digitized. Figure 50 (a) depicts the essential components of a digital fluoroscopy system. The center of the system is a computer that controls the firing of the X-ray tube, imaging acquisition via the camera, and imaging processing such as subtraction.

## 1. Digital Subtraction Angiography (DSA)

The conventional film subtraction technique has been found useful in eliminating the background noise and enhancing the image of the object of interest. An important example is temporal subtraction angiography in which the image before the injection of a contrast medium is subtracted from the image obtained shortly following the injection. This procedure can be readily performed digitally. In fact, when this idea was first introduced, there was the expectation that digital processing with inherent better signal-to-noise ratio would allow intravenous injection in subtraction angiography as opposed to intra-arterial injection. Venous injection is preferred over arterial injection because the procedure is safer (the veins are low-pressure vessels). This optimism is based on the fact that digital systems are typically more immune from noise, larger signal amplification may be possible in the video processing chain, and signal processing algorithms such as contrast enhancement through windowing and filtering can be implemented. It was thus envisioned that these improvements might allow the detection of even minute concentrations of contrast medium in arteries following intravenous injection. The less risky nature of intravenous injection in certain cases makes hospitalization unnecessary. Although this optimism has not materialized up to now because more noises are introduced into the image by the additional components in the hardware, and the quantum mottle noise, defined in Section V of this chapter, becomes more severe following subtraction, DSA still possesses a number of advantages over film subtraction angiography (High, 1988): (1) DSA allows further image processing; (2) because of better sensitivity, smaller concentrations of contrast medium may be used, with a reduction in the probability of patient allergic reaction, the injection can be performed with a smaller catheter, lessening the risk associated with puncturing an artery, and DAS may be performed as an outpatient procedure; (3) the problem caused by patient motion is reduced since there is less discomfort associated with the injection of smaller amount of the contrast agent; (4) it is less costly in the long run due to saving in material costs (mainly film); and (5) there is improvement in speed and efficiency.

However DSA is not without its limitations. It suffers from poorer resolution and smaller field of view compared to film subtraction angiography and, it is not possible to produce multiple views simultaneously.

The major components of a DSA system are similar to those of a digital fluoroscopy system with the exception that the video signal from the video cam-

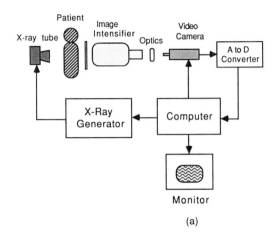

(a)

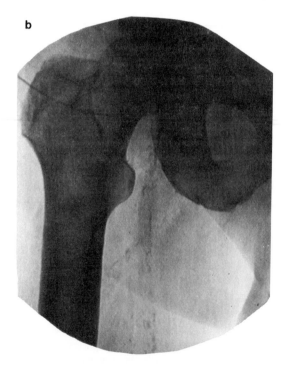

**Figure 50**   (a) Digital fluoroscopic system. Images (b), (c), and (d) are respectively digital arteriograms of iliac artery before and after intra-arterial injection of contrast medium and digitally subtracted image. (Courtesy of Philips Medical Systems, Shelton, Connecticut).

c

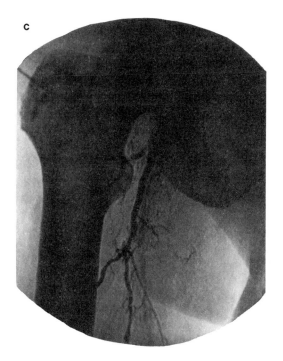

d

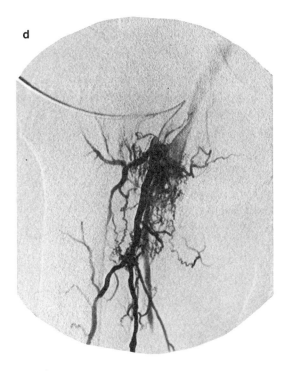

**Figure 50**   (*continued*)

era is typically logarithmically amplified before digitizing. The reason is two fold: (1) Logarithmic compression reduces the dynamic range of the input signal so as to make the blood vessel in the subtracted image appear more uniform; and (2) the logarithmically compressed signal is directly related to the concentration of the contrast agent in the vessel.

Figure 50(d) shows the subtracted image of iliac artery obtained with digital angiography following intra-arterial injection; (b) and (c) are images before and after the injection of contrast medium.

## 2.  Dual Energy Subtraction

The dual energy subtraction technique was pioneered by Alvarez and Macovski (1976) (Brody, 1984). [Here only a qualitative discussion is given. Mathematically vigorous treatment of the subject can be found elsewhere (Huang, 1987).] It uses the principle that the attenuation coefficients of different materials in regions where the attenuation coefficient drops smoothly as the energy is increased (Fig. 6) vary differently as X-ray energy is changed. For instance, the soft tissue decreases more slowly than bone or iodine. The approach can be better understood by considering the following example. Suppose we have a phantom as shown in Fig. 51(a) illuminated by an X-ray beam, say at 80 keV. The image is collected by a digital X-ray system. The gray scale of the display set by the system is shown in Fig. 51(b). The phantom is then illuminated with an X-ray of higher energy, say 120 keV, and in displaying the image the gray scale is changed to what is shown in Fig. 51(c), where the gray level for the soft tissue matches that for the soft tissue at 80 keV. It then becomes quite apparent that, following subtraction of one image from another, the remaining image contains no information about the soft tissue. Similarly, and image containing only soft tissue can also be obtained.

Dual energy subtraction solves the problem of patient motion frequently associated with temporal subtraction technique. However, it also has problems of its own. Among the most important important ones are that (1) it smears or reduces the difference in gray scale between adjacent regions, and (2) the complexity of the X-ray generator is increased (Brody, 1984).

## 3.  K-edge Subtraction

K-edge subtraction is a variation of dual energy subtraction (Mistretta et al., 1973), which utilizes the abrupt changes of attenuation coefficients of contrast agents such as iodine (Fig. 6) near their K-edges. The K-edge absorption for iodine occurs at 33.2 keV. If two images are taken with monochromatic X-ray beams at energies of 33 and 34 keV and a subtraction is performed on the two images, the difference in gray level between structures containing iodine and the surrounding tissues in the subtracted image should be maximized, since attenuation coefficients of tissues including bone are nearly identical at these two energies. However, several problems have to be overcome before this technology can be used clinically. The most important problem has been that the X-rays produced by conventional X-ray tubes are polychromatic. To generate monochromatic beams with sufficient intensity, expensive synchrotron radiation sources have to be used. Alternatively, filters can be used to select the energies of interest from X-ray spectra produced by conventional X-ray generators. For instance, iodine filters can be used to select energy spectra slightly below 33.2 keV and cerium filters can be used to select spectra slightly above 33.2 keV. Unfortunately, to be

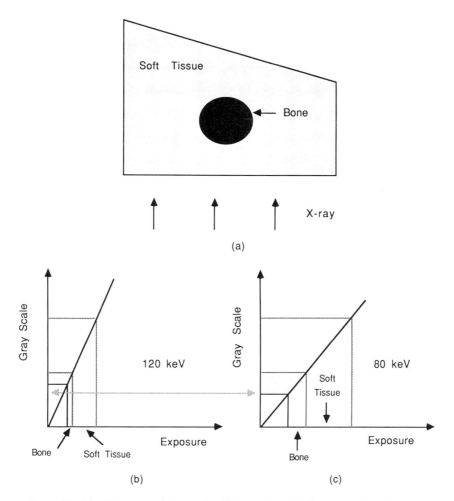

**Figure 51**   Idealistic example is used to illustrate how dual-energy subtraction works. (a) Tissue phantom consisting of bone enclosed by soft tissue. (b) Gray scale versus exposure at 120 keV. (c) Gray scale versus exposure at 80 keV. Exposures due to different X-ray energies vary but gray level of soft tissue can be adjusted to match.

able to achieve a narrow radiation band, very thick filters have to be used. This creates the problem that the X-ray tube load has to be high (Brody, 1984).

## B.   3-D Reconstruction

If multiple slices of cross-sectional CT images are obtained and stored in computer memory in an organized format (attenuation coefficient at a certain voxel and the coordinates of the voxel), it seems intuitive that the image of a slice along any plane (e.g., coronal or saggital plane) or the image of the object in 3-D can be reconstructed (Udupa, 1983; Herman, 1988). There are now a number of companies that specialize in 3-D imaging software. This software can be categorized into two different approaches: surface and volume rendering. Figure 52 shows the 3-D image of a human skull reconstructed from multiple CT slices.

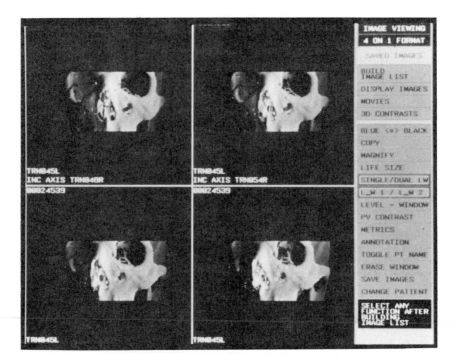

**Figure 52**   Three-dimensional surface rendering of a human skull from multiple slices of CT images. (Courtesy of Cemax, Inc, Santa Clara, California).

Which 3-D rendering technique is better depends on the information that needs to be extracted and the clinical applications (Udupa and Herman, 1989). Software is also available for manipulating the image in various ways, such as rotation and slicing.

The advantage of a 3-D image is that a global view of the structure is obtained. There is no need for the physicians to reconstruct the object from 2-D views in their mind. The disadvantages are (1) More and thinner slices (1 to 3-mm slices as opposed to 10 mm) have to be taken. This means that patient radiation exposure is increased. (2) Smoothing algorithms are used to improve the image but they may also smear out diagnostically important information. (3) More operator time and skill is needed. (4) Image can be distorted by motion artifact. It is because of these problems that the significance of 3-D imaging in medicine is uncertain, although it has been found useful for planning orthopedic and cranial surgery and for planning cancer therapy (Vannier, 1987, and Freiherr, 1987). Although 3-D imaging is technically sophisticated, at present, it is viewed by physicians basically as an adjunct to other modalities, since it does not provide any new information but rather presents the known data in a different and more understandable way.

Three dimensional imaging is of emerging interest not only in X-ray but also in other imaging modalities such as ultrasound, magnetic resonance imaging, and nuclear imaging. Various technical and clinical issues in this technology are still being examined and debated (Herman, 1990).

## C.  Dynamic Spatial Reconstructor (DSR)

Computed tomography presents accurate anatomic images of cross sections of the body. Because of the scan time involved, it is impossible to obtain a real-time three-dimensional image of the object by conventional CT scanners. This capability is important for imaging the heart. Two different types of systems, one at the Mayo Clinic (Robb *et al.*, 1983) and one at the University of California–San Francisco (Boyd *et al.*, 1983), have been developed to meet this need. A machine developed at Mayo and given the name "dynamic spatial reconstructor," which can reconstruct complete dynamic three-dimensional images of moving structures in the body from multiplanar X-ray projections, is discussed in this section; the UC–San Francisco machine, which uses electron beam technology, will be discussed in the next section.

Figure 53 shows the block diagram of an early version of the system. The central part of the system is a digital fluoroscopic system, shown at the bottom of the figure. The object to be imaged is positioned in the X-ray field and rotated under computer control. The fluoroscopic image at each successive angle of view is scanned by the video camera 60 frames/second. The video image is made up of about 100 to 200 horizontal scan lines. Measurements of the transmitted X-ray intensity are achieved by digitizing each scan line with the video digitizer. To image a moving object such as the heart, the cardiac, and respiratory cycles, the X-ray tube, the rotation motor, and the computer are all under the control of the master oscillator labeled SYNC in Fig. 53. Since each video image is the projection image of the entire chest, the 60 images accumulated at one angle in one second will include the whole cycle of a heart beat. When the object is rotated to another angle, the temporal relationships among the cardiac pacesetting pulse, respiratory motion, and the video scans are exactly maintained by 60 per second oscillator pulses. For instance, let us assume that the 36th video image represents the projection image of the heart at the peak of the systole and the 80th video line represents the projection profile for a section of the heart through the mid-por-

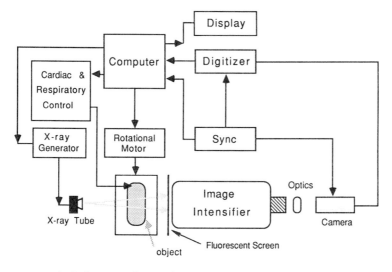

**Figure 53**  Block diagram of an early DSR.

tion of the heart. So, from all of the 80th video lines of the 36th video images collected at different angles, the CT reconstruction algorithms can be used to reconstruct the cross-sectional image. Then from these cross-sectional images, a three-dimensional image of the heart at systole can be reconstructed.

This early system used one X-ray source and one detector. The object to be imaged was rotated to different angles. The more updated version, shown in Fig. 54(a), is designed to hold 28 X-ray generators equally spaced in a semicircle and 28 video cameras arranged in the opposing semicircle. The patient stays stationary while the gantry containing the X-ray tubes and cameras is rotated about the patient continuously at 15 rev/min or 1.5° per 1/60 sec. The image of the structure, illuminated by a wide X-ray beam, is projected on a curved fluorescent screen and captured by the camera. Any portion of the body can be scanned by appropriately positioning the table. One complete volume (up to 240 cross sections) can be obtained from 28 views recorded in 1/100 of a second and repeated 60 times/sec. Special-purpose hardware and software have been designed to perform reconstructions of selected single cross sections on-line, and to provide efficient off-line calculations of all dynamic volume images generated in a DSR scan. Multi-oriented sections through the scanned volumes can be displayed, as well as full 3-D representations of the region scanned.

Figure 54(b) shows the reconstructed images of a dog's heart obtained with this system.

## D. Imatron or Fastrac Electron Beam CT

This type of system is now commercially available. The mechanically moved gantry in conventional CT is too slow to image a fast-moving structure like a heart. To shorten the scanning time, either multiple sources and detectors, as in DSR, or other means to sweep the X-ray beam will have to be used. Scanning electron-beam technique is a method belonging to the latter category. A schematic diagram indicating the basic components of such a system is shown in Fig. 55(a). An electron beam is focused by a set of focusing coils and deflected to the target consisting of four tungsten rings spanning 210° by a set of deflection coils [Fig. 55(a) and (b)]. Opposite the tungsten rings is an array of luminescent crystal silicon photodiodes, spanning an arc of 216°, acting as detectors. During scanning, which takes approximately 50 msec/scan, the electron beam is swept by the deflection coil. As the electron beam strikes the tungsten target, X-rays are generated and are collimated into a fan beam by the collimator. Four different imaging slices can be selected by striking different tungsten rings. It can achieve an impressive maximal scanning rate of 17 scans/sec. This type of scanner has been found useful in diagnosing a number of cardiac problems (Sethna *et al.*, 1987).

## E. Other Developments

The discussions given in preceding sections are not intended to be exhaustive. There are many other important and interesting developments that are not included in this book. Line scan digital radiography (Barnes, 1985), flying spot scanning (Huang, 1987), Compton backscatter dynamic radiography (Towe and Jacobs, 1981, and McInerney *et al.*, 1984), picture archiving and communication

a  *DYNAMIC   SPATIAL   RECONSTRUCTION   SYSTEM*

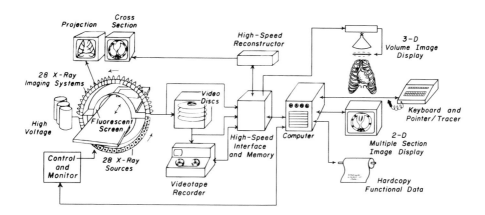

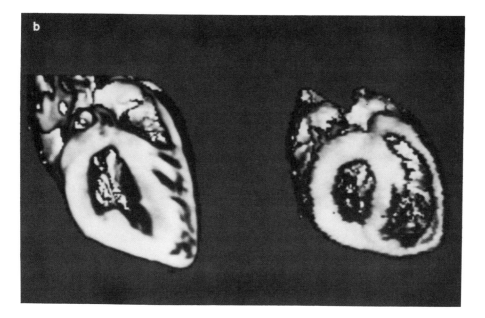

**Figure 54**   (a) Block diagram of a more recent DSR including scanner at left, recording and computational equipment in center, and multidimensional display and analysis devices at right. (b) 3-D volume rendered images of dog heart obtained with DSR. Heart is mathematically dissected during rendering to show long-axis view at left and short-axis view at right of left ventricular chamber. (Courtesy of Richard A. Robb, Mayo Clinic, Rochester, Minnesota).

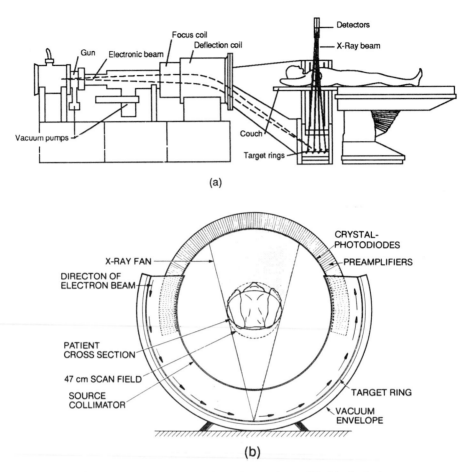

**Figure 55** (a) Schematic diagram of an electron beam CT; (b) physical construction of its gantry. (Courtesy of Picker International Inc, Cleveland, Ohio and Imatron Inc, South San Francisco, California).

systems (PACS) (Huang, 1987), and dual-energy bone densitometry (Johnson *et al.*, 1991) are just a few examples. Readers may refer to the list of bibliography for further information.

# V.   X-ray Image Characteristics

Perhaps the most important goal in medical imaging is to achieve "optimum image quality" such that the "best" clinical diagnosis can be obtained. The definition of optimum image quality is elusive because it depends on image characteristics, the detection task used in a specific clinical diagnosis, and the visual perception model for the human eye–brain system. A number of indices based on measurable image characteristics and simple models of detection have been proposed to describe image quality. Typical examples are those using the concepts of signal-to-noise ratio based on statistical theory of detection (Wagner *et al.*, 1985). These image quality indices work well for simple detection tasks using images obtained from simple object configurations. In clinical diagnosis the detection tasks are often too complicated to match the theoretical model. Also, the lack of complete understanding of the psychophysiological function of the human eye–brain system is another major obstacle in the development of a universal image quality index. For these reasons, the best current method for image evaluation is the use of observer performance study and receiver operating curve (ROC) analysis, which will be described in a later section.

Despite the difficulties in our full understanding of image quality as described above, descriptions of image characteristics based on measurable parameters are strong indicators of the quality of X-ray images. The three most important image characteristics are spatial resolution, noise, and contrast, which are described below.

## A. Spatial Resolution

Spatial resolution of an imaging system is defined as the ability of the system to differentiate fine structures. A commonly used measure of spatial resolution is the minimum separation between two point or line objects that can be resolved by the imaging system. In X-ray imaging, this qualitative measure of spatial resolution is usually obtained using a bar phantom, shown in Fig. 56. The bar phantom consists of groups of lead strips with different widths; the separation of lead strips in each group is equal to the width of the strips. An image of the phantom shows groups of band or line pairs with alternating dark and bright intensities or spatial frequencies, such as 4 lines/mm, 6 lines/mm, etc. The spatial resolution of the imaging system is determined by the highest spatial frequency of line pairs that can be resolved. A good X-ray imaging system can resolve 6 or 8 lines/mm while a CT can resolve about 2 lines/mm.

The measure of spatial resolution by resolving line pairs is qualitative and subjective. A more rigorous and complete description of spatial resolution can be derived by treating the imaging process as a linear system (Metz and Doi, 1979). The following are a number of descriptors that have been used.

### 1. Point Spread Function

The two-dimensional point spread function $psf(\mathbf{r})$ of a linear and shift-invariant system is defined as the normalized point response function $prf'(\mathbf{r})$, that is, the response of the system to an input of point impulse. The point spread function is

**Figure 56** Bar phantom for determining spatial resolution of X-ray systems.

normalized in the following sense:

$$\text{psf}(\mathbf{r}) = \frac{\text{prf}'(\mathbf{r})}{\iint \text{prf}'(\mathbf{r}) \, d\mathbf{r}} \tag{1-20}$$

where $\mathbf{r} = (x, y)$. In general, a narrower psf($\mathbf{r}$) represents better spatial resolution. Moreover, the shape of psf($\mathbf{r}$) provides complete information about the spatial resolution characteristics of the system in the absence of noise.

The prf($\mathbf{r}$) of an X-ray imaging system can be obtained experimentally by imaging a small hole bored through a lead sheet. By using the H–D or characteristic curve of the film, the optical density distribution of the recorded image on a processed film is converted to the exposure distribution and the prf($\mathbf{r}$). To obtain accurate measurement, the size of the hole has to be small compared to that of the prf($\mathbf{r}$). This requirement imposes practical difficulties in the direct measurement of the prf($\mathbf{r}$).

## 2.   Line Spread Function

Line spread function is defined as the one-dimensional integral of the point spread function along a given direction, that is,

$$\text{lsf}(x) = \int \text{psf}(\mathbf{r}) \, dy = \int \text{psf}(x, y) \, dy \tag{1-21}$$

where $\mathbf{r} = (x, y)$. For a linear and shift-invariant system, such as the X-ray imaging system, with isotropic psf($\mathbf{r}$), the one-dimensional lsf($x$) is the same in any orientation and uniquely specifies the system response. Experimentally, the lsf($x$) can be measured from imaging a narrow slit between two lead sheets. The experimental measurement is much easier compared to that of the point spread function. The lsf($x$) can be derived from the one-dimensional profile through the exposure distribution in the direction perpendicular to the length of the slit. Figure 57 shows the lsf($x$) of two X-ray imaging systems with different screen–film combinations indicating the difference in spatial resolution. Similar to the psf($\mathbf{r}$), the shape of the lsf($x$) provides full information about the spatial resolution characteristics of the imaging system.

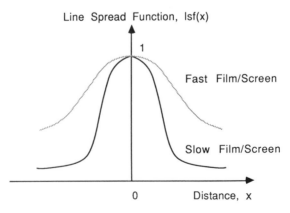

**Figure 57**   Line spread functions for fast and slow film/screen combinations of conventional X-ray radiography.

## 3. Edge Spread Function

The edge spread function esf(**x**) is related to the line spread function by

$$\text{lsf}(x) = \frac{d}{dx}\,\text{esf}(x) \tag{1-22}$$

The $\text{esf}(x)$ can be measured easily using the experimental arrangement shown in Fig. 58. It is often used as an intermediate means to derive the $\text{lsf}(x)$ of the imaging system. Also, the $\text{esf}(x)$ is useful in characterizing edges in an image for various applications.

## 4. System Transfer Function

The two-dimensional system transfer function $\text{STF}(\rho, \xi)$, defined as the spatial Fourier transform of the two-dimensional psf(**r**), has been found useful in imaging system analysis. In particular, the modulation transfer function $\text{MTF}(\rho)$, defined as the magnitude of the spatial Fourier transform of the $\text{lsf}(x)$, that is,

$$\text{MTF}(\rho) = \left| F_1[\text{lsf}(x)] \right| \tag{1-23}$$

is widely used in characterizing the imaging system. The modulation transfer functions of the two imaging systems characterized by the two $\text{lsf}(x)$ in Fig. 57 are shown in Fig. 59.

The $\text{MTF}(\rho)$ provides information about how the system modulates the spatial frequency content of the input distribution. As shown in Fig. 59, most X-ray imaging systems act like low pass filters in allowing low spatial frequencies to pass through and suppressing high spatial frequencies. This results in blurring of fine structures contained in the input distribution. From Figs. 57 and 59, it can be seen that the imaging system with the poorer spatial resolution, or wider $\text{lsf}(x)$, gives the narrower $\text{MTF}(\rho)$.

The $\text{MTF}(\rho)$ is useful in transfer function analysis because it is often easier to analyze the imaging process in the spatial frequency domain. For example, the

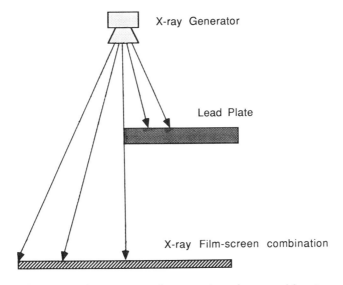

**Figure 58** Experimental arrangement for measuring edge spread function.

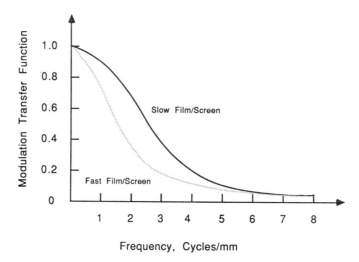

**Figure 59**   Modulation transfer functions for fast and slow film/screen combinations of conventional X-ray radiography.

image i($\mathbf{r}$) of an object of o($\mathbf{r}$) obtained using an imaging system characterized by the psf($\mathbf{r}$) of the system is given in the spatial domain by

$$i(\mathbf{r}) = o(\mathbf{r}) \text{ ** } psf(\mathbf{r}) \tag{1-24}$$

where $\mathbf{r} = (x, y)$ and ** is the two-dimensional convolution operation. Note here that psf($\mathbf{r}$) is equivalent to $\mu$ in Eq. (1-15). In the spatial frequency domain, the image spectrum I($\rho, \xi$), or the two-dimensional Fourier transfrom i($\mathbf{r}$), is given by

$$I(\rho, \xi) = O(\rho, \xi) \cdot H(\rho, \xi) \tag{1-25}$$

where $O(\rho, \xi)$ is the object spectrum, $H(\rho, \xi)$ is the system transfer function, MTF$(\rho, \xi) = |H(\rho, \xi)|$ and · is the multiplication operation. By using fast Fourier transform algorithms, image analysis using Eq. (1-25) has been found computationally more efficient compared with using Eq. (1-24), which involves the convolution operation. Also, imaging analysis in the spatial frequency domain often provides insights not found in the spatial domain.

## B.   Image Noise

The most important sources of image noise in an X-ray imaging system are the statistical fluctuations in X-ray photon detection and the grain size in the screen–film combination (Rossman, 1963; Doi, 1969). The image noise in an X-ray image is often described by the quantum mottle (Rossman, 1963). A complete quantitative decription of X-ray image noise is the use of Wiener power spectrum (Doi, 1969), which provides measures of the magnitude and texture of the image noise. For example, for images that exhibit coarser grain structures, the Wiener power spectrum will be narrower in shape. The magnitude of the noise fluctuation is given by the integral of the Wiener power spectrum. Image noise can be removed by averaging.

## C. Image Contrast

Image contrast is another important factor affecting image quality. Suppose the signal and background intensities are given by $S$ and $b$, respectively. Contrast can be defined as

$$C = \frac{S}{S + b} \quad \text{or} \quad \left| \frac{S - b}{S} \right| \tag{1-26}$$

Since most detection tasks impove with increased contrast, it is desirable to have an imaging system that generates higher image contrast for the same object contrast. For example, an X-ray system with lower kVp generates higher image contrast compared with a higher kVp. However, lower kVp X-ray results in higher image noise due to photon attenuation in the tissues. The minimum contrast required to detect an object with specific size in the presence of image noise is called contrast resolution (Cohen and DiBianca, 1979).

## D. Receiver Operating Curve (ROC)

The parameters discussed above, resolution, noise, and contrast, are useful for determining the performance of an imaging device in an ideal setting. However, in a true clinical environment, the information contained in an image is much more complicated. Therefore, merely using these criteria alone cannot justify whether an imaging system is better or worse than another for diagnosing a certain disease. In other words, a system with superior resolution producing sharper images cannot guarantee that the system performs better than the other because there are many other factors involved in reaching a correct diagnosis. With the proliferation of imaging methodologies and the containment of healthcare cost on everybody's mind, the need for an objective means for assessing the performance of an imaging system is compelling. The receiver operating curve (Metz, 1978; Swets, 1979) has been used in recent years to assist in determining the diagnostic value of an imaging system. This section describes how a receiver operating curve can be obtained.

When a clinician sees a series of images that do or do not contain a certain structure, a 2 × 2 decision matrix, shown in Fig. 60, can be constructed for this person if the answer is binary, that is, either yes or no. Thus it would seem that from this decision matrix, the performance of this imaging device in detecting a

| | | True Condition: The object is there | True Condition: The object is not there |
|---|---|---|---|
| Observer Response | The object is there | True positive **TP** | False positive **FP** |
| Observer Response | The object is not there | False negative **FN** | True negative **TN** |

**Figure 60** Observer decision matrix.

certain structure for this observer can be readily determined. A few important definitions are given below.

$$\text{Accuracy} = \text{correct responses}/\text{total responses} = (N_{TP} + N_{TN})/N_T$$
$$\text{Sensitivity} = \text{true positive}/(\text{true positive} + \text{false negative})$$
$$= N_{TP}/(N_{TP} + N_{FN})$$
$$\text{Specificity} = \text{true negative}/(\text{true negative} + \text{false positive})$$
$$= N_{TN}/(N_{TN} + N_{FP})$$

Note here TPF + FNF = 1 and TNF + FPF = 1 where TPF, FNF, TNF and FPF represent, respectively, true positive fraction or sensitivity, false negative fraction, true negative fraction or specificity, and false positive fraction.

There are two problems with this approach: (1) A person's decision is subjective, and (2) a person's threshold can vary. More elaboration is needed to understand the latter. How to reach a diagnosis after a radiologist sees an image is a question still being addressed today. Many factors must be considered, including past experience, history of the patient, and cost of the next procedure. A radiologist after gathering information may feel that there is sufficient evidence or a threshold is reached for him or her to make a decision, but this threshold can be changed at will. It is known that many radiologists are willing to adjust their threshold to lower false negatives after missing a disease. To remedy the first problem, a statistically significant number of observers have to be recruited, whereas the receiver operating curve can be used to alleviate the second problem.

The ROC analysis involves having a viewer read a series of images with known abnormality. A curve as shown in Fig. 61 is then constructed from the decision matrices derived from the diagnoses made by the viewer with varying threshold. Better imaging systems or better diagnosticians have a larger area un-

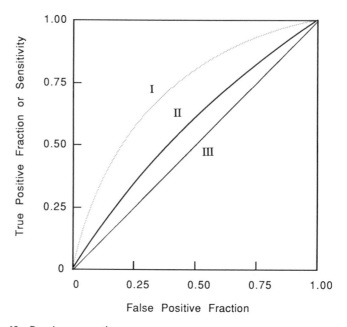

**Figure 61**   Receiver operating curve.

der the curve. The area under the curve has been shown to be equal to the accuracy of the observer or system (Swets, 1979). In Fig. 61, curve I represents a system or observer whose performance is better than that represented by system II whereas curve III represents a totally useless system or an observer who is purely guessing.

# VI.  Biological Effects of Ionizing Radiation

It has been observed for many years that ionizing radiation produces damage to the tissue (Barnett, 1977; Krasin and Wagner, 1988). One stunning evidence is the epidemiological study of people exposed in utero at the time of the atomic bombing of Hiroshima. These people and a control group have been followed through childhood to adulthood. Data obtained show that the incidence of leukemia and mental retardation of the exposed group correlates well with proximity to the bomb, thus with the level of exposure. In this section, we will review briefly the various types of biological effects of ionizing radiation and their relationship to radiation exposure parameters, such as time, area, and dose.

## A.  Determinants of Biological Effects

### 1.  Threshold

Ideally, to be able to determine what level of exposure is acceptable, it is desirable to establish a quantitative dose level above which damage to tissues occurs and below which there is no response, or the threshold dose. Unfortunately, this level either does not exist or is extremely difficult to determine.

### 2.  Exposure Time

The severeness of the biological effects depends on the duration of exposure. Since a considerable degree of recovery occurs from the radiation damage, a given dose will produce less effect if divided than if it were given in a single exposure.

### 3.  Exposure Area

The larger the body area exposed, the greater the damage. This is why, in radiation therapy, the radiation is directed at the part of the body to be treated while the rest of the body is safely shielded from the radiation.

### 4.  Variation in Species and Individual Sensitivity

There is wide variation in the radiosensitivity of various species. Generally, plants and microorganisms are much less susceptible to radiation damage than mammals. Within the same species, individuals vary in sensitivity. For this reason, the lethal dose for each species is expressed in statistical terms, usually as the LD 50/30 for that species, or the dose required to kill 50% of the individuals in a large population in a 30-day period. For humans, the LD 50/30 is estimated to be around 450 rads for whole body irradiation.

**Table III**
Radiation dose received by the patient
in several X-ray procedures.

| X-ray procedure per exposure | Exposure mR |
| --- | --- |
| Chest | 20 |
| Brain | 250 |
| Abdomen | 550 |
| Dental | 650 |
| Breast | 54 |
| Xeromammography | 200 |
| CT/slice | 1000 |

## 5.   Variation in Cell Sensitivity

Within the same individual, the most sensitive to radiation damage are those cells that are rapidly dividing. Furthermore, nonspecialized cells are more sensitive than specialized cells. Based on these factors, it is easy to understand that white blood cells are the most sensitive, followed by immature red cells. Next are the epithelial cells. Muscle and nerve cells generally are less sensitive.

## B.   Short Term and Long Term Effects

If a dose of radiation of unusually large quantity, generally over 100 rads, is delivered to the body in a very short time, the biological effects that occur within a period of hours or days after the irradiation are referred to as short term effects. The signs and symptoms that comprise these short term effects are known as the acute radiation syndrome, which is characterized initially by nausea, vomiting, and malaise, and subsequently by more severe disorders like fever, shock, or even death.

In diagnostic X-ray, long term and more subtle effects caused by low dose radiation that may manifest themselves years after the exposure are more important. Of the possible effects, (1) carcinogenic effects and (2) genetic effects have received the most attention. There is mounting evidence now to indicate that these effects can be induced by diagnostic level X-ray. It is for this reason that the National Council on Radiation Protection and Measurement (Bethesda, Maryland) established in 1958 that the maximal allowable dose per person per year is 5 R or an average 0.2 R per working day per person. Table III lists the dose received by the patient during various X-ray procedures.

## Problems

1. In Fig. 6 of the text it is seen that X-ray attenuation coefficient in soft tissue decreases as photon energy increases for photon energy greater than 20 keV. Is this true for photon energy levels less than 20 keV? Explain.
2. What is the most important scattering process in soft tissues for X-ray greater than 60 keV? Explain.

**3.** Muscle and bone are arranged as shown. The densities for muscle and bone are 1 gm/cm$^3$ and 1.8 gm/cm$^3$ respectively. The mass attenuation coefficients are

| E | Muscle | Bone |
|---|---|---|
| 60 keV | 0.20 | 0.27 |
| 1 meV | 0.07 | 0.07 |

Compare the intensity of the attenuated X-ray beam that has passed through bone and muscle with just the muscle at the two energies.

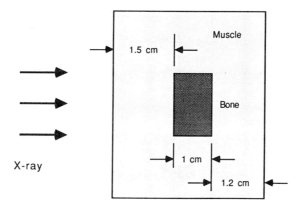

**4.** Two intensifying screen–film systems have the MTF curves as shown. **(a)** Which screen is better if the resolution of the eyes is between 2 to 4 lines/mm? **(b)** Which screen is desirable for recording very small structures?

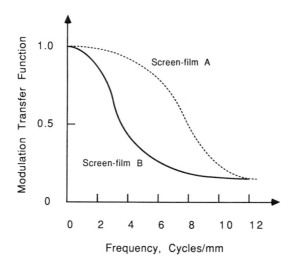

**5.** For the films with characteristic curves as shown: **(a)** Which film has a better contrast? **(b)** How does the speed of film A compare with film B? Assume that the base plus fog density is negligible.

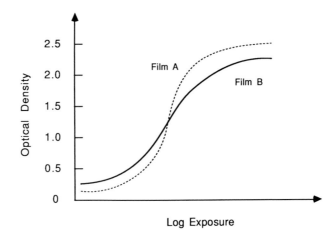

6. A double-emulsion film as shown in Fig. 22 is usually used to record X-ray images. Discuss how the image contrast can be increased by using a double-emulsion film over a single-emulsion film assuming that X-ray attenuation and filtration in the base and intensifying screen can be ignored; that is, the intensity of X-ray reaching the intensifying screen in the back is the same as the intensity at the front screen.

7. A test pattern as shown, which is placed above the focal plane, is being imaged by a linear conventional tomography. **(a)** Draw an image of the pattern. **(b)** Discuss how slice thickness in focus is related to the distance traveled by the source.

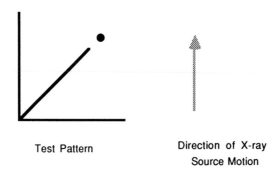

Test Pattern                    Direction of X-ray
                                 Source Motion

8. Find the modulation transfer function if the impulse response of a screen–film combination is as shown for $a = 100$ cm$^{-1}$ and $a = 10$ cm$^{-1}$. If a value of 0.1 of MTF is needed to resolve two lines, what are the spatial resolutions of the two systems?

Line Spread Function

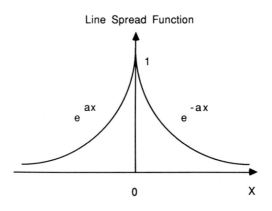

**9.** It has been shown that the line spread functions for the X-ray film and intensifying screen can be approximated by the following functions: **(a)** Film: $LSF(x) = (1/2L)$ $\exp(-|x|/L)$ where $L$ is a constant for a particular emulsion. **(b)** Screen:

$$LSF(x) = \frac{\beta m}{4\pi} \ln\left[\frac{x^2 + d_s^2}{x^2 + (d_s - d_1)^2}\right]$$

where $d_1$ = thickness of phosphor
$\quad\quad d_s$ = thickness of screen
$\quad\quad \beta$ = attenuation coefficient of phosphor
$\quad\quad m$ = number of optical photons emitted by absorbed X-ray photon

Plot these functions and then find and plot the modulation transfer functions, MTF $(\rho)$, for film, screen, and film–screen combination. Note that modulation transfer function has to be normalized to this value at $\rho = 0$.

**10.** Find and plot the projection function along $x$-direction of an object as shown. The object is a homogeneous disk with an X-ray attenuation coefficient of 0.1 cm$^{-1}$ of 1 cm radius.

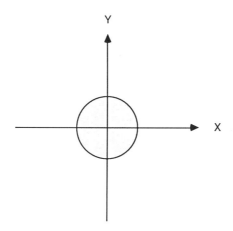

**11.** Given $\delta[f(x)] = \sum_i |df/dx|_{x=x_i}^{-1} \delta(x - x_i)$ where $f(x_i) = 0$ and $\delta(x)$ is a delta function, show that the back projection function $\Omega(r)$ is given by

$$\Omega(r) = \beta(r) ** (1/\pi r)$$

where $\beta(r)$ is the linear attenuation coefficient at location $r$ of an object to be reconstructed.

**12.** Discuss the possible problems that may be encountered in estimating blood vessel diameter in X-ray angiography. Suppose you are given the task to design an experimental protocol to evaluate the accuracy of angiography in determining blood vessel diameter. How would you do it?

## References and Further Reading

Alvarez, R. E. and Macovski, A. (1976). *Phys. Med. Biol.* **21,** 733.

Arnold, B. A. (1979). Physical characteristics of screen–film combination. *In* "The Physics of Medical Imaging." (A. G. Haus, ed.), *Am. Inst. Phys.* New York.

Bange, F. (1988). X-ray attenuation in matter. *In* "Encyclopedia of Medical Devices and Instrumentation." (J. G. Webster, ed.), John Wiley, New York.

Barnes, G. T. (1985). *Radiol.* **154,** 801.

Barnett, M. H. (1977). "The Biological Effects of Ionizing Radiation: An Overview." HEW Public. (FDA) 77-8004, Washington, D.C.

Barrett, H. H., and Swindell, W. (1981). "Radiological Imaging: The Theory of Image Formation, Detection, and Processing." Academic Press, New York.

Bassett, L., and Gold, R. H. (1987). "Breast Cancer Detection: Mammography and Other Methods in Breast Imaging, 2E" Grune and Stratton, Orlando.

Bettmann, M. A., and Morris, T. W. (1986). *Radiol. Clin, N. Am.* **24,** 347.

Boyd, D. P., and Lipton, M. J. (1983). *Proc. IEEE* **71,** 298.

Brody, W. R. (1984). "Digital Radiography." Raven Press, New York.

Cohen, G., and DiBianca, F. A. (1979). *J. Comput. Assist. Tomogr.* **3,** 189.

Cormack, A. M. (1963). *J. App. Phys.* **34,** 2722.

Curry, T. S., III, Dowden, J. F., and Murry, R. C., Jr. (1990). "Christensen's Introduction to the Physics of Diagnostic Radiology, 4th ed." Lea and Febiger, Philadelphia.

des Plantes, B. Z. (1934). "Plantinigraphie en Substratie Roentgenographische Differentiatiemethoden." Thesis, University of Utrecht, The Netherlands.

Doi, K. (1969). Wiener spectrum analysis of quantum statistical fluctuation and other noise sources in radiography. *In* "Television in Diagnostic Radiology." (R. D. Mosley, Jr. and J. H. Rust, eds.) Aesculapius.

Freiherr, G. (1987). *Diagnostic Imaging* **9,** 190.

Herman, G. T. (1988). *J. Comput. Assist. Tomog.* **12,** 450.

Herman, G. T. (1990). *IEEE. Eng. Med. Biol. Mag.* **9,** 15.

High, M. (1988). Digital Angiography. *In* "Encyclopedia of Medical Devices and Instrumentation." (J. G. Webster, ed.) John Wiley, New York.

Hirose, A., and Lonngren, K. E. (1985). "Introduction to Wave Phenomena." John Wiley, New York.

Hounsfield, G. N. (1973). *Br. J. Radiol.* **46,** 1016.

Hobbie, R. K. (1978). "Intermediate Physics for Medicine and Biology." John Wiley, New York.

Huang, H. K. (1987). "Elements of Digital Radiology." Prentice Hall, Englewood Cliffs, New Jersey.

Johnson, C. C., Jr., Slemenda, C. W., and Melton, L. J., III. (1991). *N. Engl. J. Med.* **324,** 1105.

Keller, F. S., and Routh, W. I. (1990). *Diagnostic Imaging,* September, 76.

Kopans, D. B. (1989). "Breast Imaging." J. B. Lippencott, St. Louis.

Krasin, F., and Wagner, H., Jr. (1988). Biological Effects of Ionizing Radiation. *In* "Encyclopedia of Medical Devices and Instrumentation." (J. G. Webster, ed.), John Wiley, New York.

Kruger, R., and Riederer, S. (1984). "Basic Concepts of Digital Subtraction Angiography." G. K. Hall, Boston, Massachusetts.

Laughlin, J. S. (1988). History of Medical Physics. *In* "Encyclopedia of Biomedical Engineering and Instrumentation." (J. G. Webster, ed.), John Wiley, New York.

McInerney, J. J., Herr, M. D., Kenney, E. S., Copenhaver, G. L., and Zelis, R. (1984). *Invest. Radiol.* **19,** 385.

Metz, C. E., and Doi, K. (1979). *Phys. Med. Biol.* **24,** 1079.

Metz, C. E., Starr, S. S., and Lusted, L. B. (1978). *Radiology* **121,** 337.

Mistretta, C. A., Ort, M. G., Kelcz, F., Camerom, J. R., Siedband, M. P., and Crummy, A. B. (1973). *Invest. Radiol.* **8,** 402.

Radon, J. (1917). *Ber. Verh. Sachs. Acad. Wiss.* **69,** 262.

Robb, R. A., Hoffman, E. A., Sinak, L. J., Harris, L. D., and Ritman, E. L. (1983). *Proc. IEEE* **71,** 308.

Rossman, K. (1963). *Am. J. Roent.* **90,** 863.

Schoenberger, S. B., Sutherland, C. M., and Robinson, A. E. (1988). *Radiol.* **168,** 665.

Sethna, D. H., Bateman, T. M., Whiting, J. S., and Forrester, J. S. (1987). *Am. J. Card. Imag. 1,* **1,** 18.

Swenberg, C. E., and Conklin, J. J. eds. (1988). "Imaging Techniques in Biology and Medicine." Academic Press, San Diego.

Swets, J. A. (1979). *Radiology* **14,** 109.

Thomas, C. (1988). Screen–film System. *In* "Encyclopedia of Biomedical Engineering and Instrumentation." (J. G. Webster, ed.), John Wiley, New York.

Towe, B. C., and Jacobs, A. M. (1981). *IEEE Trans. Biomed. Eng.* **28,** 717.

Udupa, J. K. (1983). *Proc. IEEE,* **71,** 420.

Udupa, J. K., and Herman, G. T. (1989). *Comm. Assoc. Comput. Machinery* **32,** 1364.

Vannier, M. W. (1987). *Diagnostic Imaging,* **9,** 11, 206.

Wagner, R. F., and Brown, D. G. (1985). *Phys. Med. Biol.* **30,** 489.

Webb, S. (1988). "The Physics of Medical Imaging." Adam Hilger, Bristol, U.K., and Philadelphia.

Wells, P. N. T. (1982). "Scientific Basis of Medical Imaging." Churchill Livingstone, Edingburgh, U.K., and New York.

Wilks, R. (1981). "Principles of Radiological Physics." Churchill Livingstone, Edingburgh, U.K., and New York.

Zelch, J. V. (1989). *Diagnostic Imaging* November, 67.

# CHAPTER 2

# Ultrasound

The potential of ultrasound as an imaging modality was realized as early as the late 1940s when several different groups around the world, borrowing sonar technology used in World War II, started exploring the diagnostic capabilities of ultrasound (Goldberg and Kimmelman, 1988). Among them were Wild and Reid in Minnesota, Howry's group at Denver, and Fry's group at University of Illinois in the United States, Uchida, Tanaka, and Kikuchi in Japan, and Dussik and Heuter in Europe. Although ultrasound had been applied to many different medical problems since then, it did not become a widely accepted diagnostic tool until the early 1970s when gray-scale ultrasound was introduced. Now it is the second most utilized diagnostic imaging modality in medicine.

## I.  Fundamentals of Acoustic Propagation

Ultrasound is a sound wave having frequency higher than 20 kHz, the upper limit of human audible range (which is from 20 Hz to 20 kHz). Since ultrasound is a wave, it transmits energy and can be described in terms of a number of wave parameters just like electromagnetic waves or radiation. For ultrasound, these parameters are pressure, density, temperature, particle displacement, etc. Unlike electromagnetic waves, however, sound requires a medium in which to travel. It cannot propagate in a vacuum. To better visualize how the sound propagates through a medium, let us model the medium as being composed of a three-dimensional structure of spheres, representing atoms or molecules, separated by perfect elastic springs representing interparticle forces.

To simplify the matter even more, we consider only a one-dimensional lattice as shown in Fig. 62(a). When a particle is pushed to a distance from its neutral position, the disturbance or force is transmitted to the adjacent particles by the springs. This sets up a chain reaction. If the driving force is oscillating back and forth, or sinusoidally, the particles respond by oscillating in the same way. The distance $U$ traveled by the particle in an acoustic propagation is called particle displacement and usually is in the order of a few tenths of a nanometer in water. The velocity of the particle oscillating back and forth is called particle velocity $u_x$, where the subscript $x$ denotes the direction of velocity, and is in the order of a few centimeters per second in water. It has to be noted that this velocity is different from the rate of energy propagating through the medium, which is defined as the phase velocity or the sound propagation velocity $c$. In water, $c = 1.5 \times 10^5$ cm/sec. As can be seen in Fig. 62(b), since the sound velocity is much greater than the particle velocity, this perturbation has already been transmitted to other particles over a much longer distance, $U'$, while the first particle moves for only a few tenths of a nanometer.

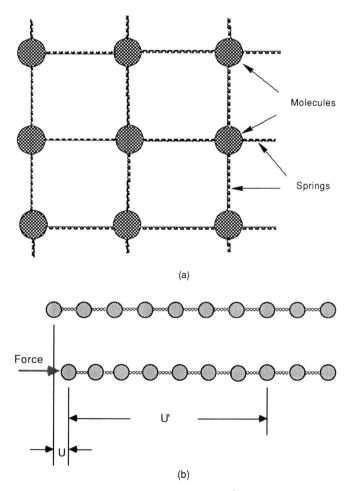

(a)

(b)

**Figure 62**   (a) One-dimensional lattice. (b) Diagram depicting difference between particle velocity and phase velocity of acoustic propagation.

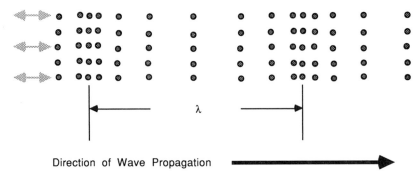

**Figure 63** One-dimensional representation of longitudinal acoustic propagation.

Regions of compression and rarefaction in a medium will be produced as shown in Fig. 63 when a sinusoidal sound wave is propagating in a liquid. The displacement of the particles is in the same direction as the direction of wave propagation. Therefore, this type of wave is called a longitudinal or compressional wave. The particle displacement is greatest in the rarefaction region, and it is smallest in the compression region. If the displacement of the particles versus distance or the displacement of a particle versus time is plotted, it can be seen that the particle moves in a sinusoidal form as was shown in Fig. 1. The sound wave has a wavelength $\lambda$ and a period $T$, which are defined in Chapter 1, Section I-A.

## A. Stress and Strain Relationship

Let us consider an incremental cube of material within a body, as shown in Fig. 64. Stress is defined as the force exerted on the incremental cube caused by other parts of the body per unit area. On a unit surface perpendicular to the $Z$ axis, the

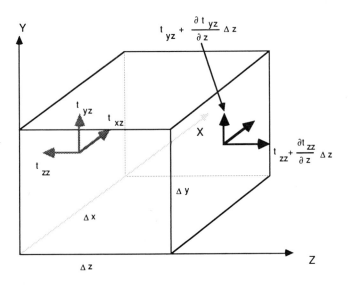

**Figure 64** Diagram illustrating the concept of stress.

stress can be separated into three components:

$t_{zz}$ is compressional stress in the $Z$ direction.
$t_{yz}$ is shear stress in the $Y$ direction.
$t_{xz}$ is shear stress in the $X$ direction.

Similarly, $t_{zy}$, $t_{yy}$, $t_{xy}$ and $t_{zx}$, $t_{yx}$, $t_{xx}$ refer to the stresses acting on the $Y$ and $X$ planes. The deformation of the cube caused by the external force can be described in terms of the following quantities or strains, which are defined as displacement per unit distance.

$$\epsilon_{zz}(\text{compressional strain}) = \frac{\frac{\partial W}{\partial z}\Delta z}{\Delta z} = \frac{\partial W}{\partial z}$$

$$\epsilon_{xz}(\text{shear strain}) = \frac{\frac{\partial U}{\partial z}\Delta z}{\partial z} = \frac{\partial U}{\partial z}$$

and so on, where $U$, $V$, and $W$ denote the displacements in $X$, $Y$, and $Z$ directions. They are functions of $(x, y, z)$. The physical meaning of compressional and shear strain is illustrated in Fig. 65(a) and (b), respectively.

Under the condition of small displacements, the stress–strain relationships are linear (Malecki, 1969).

$$t_{zz} = (\nu + 2\mu)\frac{\partial W}{\partial z} \tag{2-1}$$

$$t_{yz} = \mu\frac{\partial V}{\partial z} \tag{2-2}$$

$$t_{xz} = \mu\frac{\partial U}{\partial z} \tag{2-3}$$

where $\nu$ and $\mu$ are Lamé constants and $\mu$ is also called the shear modulus. The Lamé constants are related to the more conventional material constants by the following equations:

Young's modulus    $E = \dfrac{\mu(3\nu + 2\mu)}{\nu + \mu}$

Bulk modulus    $B = \nu + \dfrac{2\mu}{3}$

Poisson's ratio    $\xi = \dfrac{\nu}{2(\nu + \mu)}$

The definitions of these conventional elastic constants can be better understood by examining Fig. 66, where it is shown that a square bar is under tensile stress. Young's modulus is defined as the ratio of stress/strain or $t_{zz}/\epsilon_{zz}$ whereas Poisson's ratio is defined as the negative of the ratio of strain in the transverse direction to the strain in the longitudinal direction, or $-\epsilon_{yy}/\epsilon_{zz}$. The bulk modulus is the inverse of the compressibility of the material, which is defined as the negative of the change in volume per unit volume per unit change in pressure or $-(1/\sigma)(\partial\sigma/\partial p)$, where $\sigma$ denotes volume and $p$ is pressure. The definition of pressure is the normal compressional force applied on a surface per unit area of the surface. Therefore, pressure applied on a surface equals the negative of the

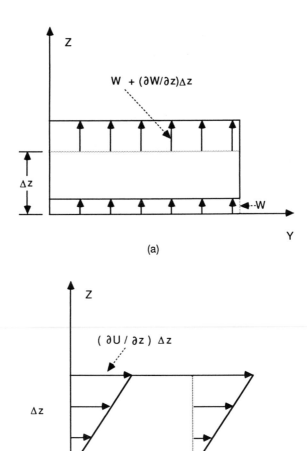

(a)

(b)

**Figure 65** (a) Compressional strain in Z-direction along Z axis. (b) Shear strain in X-direction along Z axis.

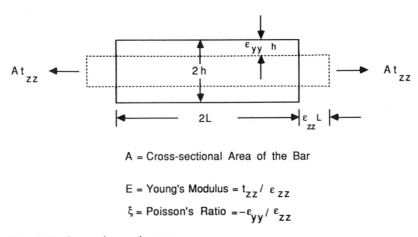

A = Cross-sectional Area of the Bar

E = Young's Modulus = $t_{zz} / \varepsilon_{zz}$

$\xi$ = Poisson's Ratio = $-\varepsilon_{yy} / \varepsilon_{zz}$

**Figure 66** Square bar under stress.

stress applied on that surface, which is the tensile force per unit area. For a surface perpendicular to the Z axis, $p = -t_{zz}$.

For a fluid, $\mu$ approaches 0, then $B \approx \nu$, and $E \approx 0$.

## B.  Equation of Motion

Consider an incremental cube as shown in Fig. 64. The equation of motion in the
Z-direction can be readily obtained by applying Newton's second law, that is, by
summing the net force applied on the cube in the Z-direction.

$$\frac{\partial t_{zz}}{\partial z} + \frac{\partial t_{zy}}{\partial y} + \frac{\partial t_{zx}}{\partial x} = \eta_0 \frac{\partial^2 W}{\partial t^2} \tag{2-4}$$

where $\eta$ is the mass density of the cube. The left-hand side of the equation is the
total force acting on the cube in the Z-direction and the right-hand side is simply
the product of the mass of the cube and the acceleration produced by the force.

### 1.  Compressional Wave

Considering only the one-dimensional case and assuming $t_{zy}$ and $t_{zx}$ are zero, that
is, there are no shear stresses, Eq. (2-4) can be simplified to

$$\frac{\partial t_{zz}}{\partial z} = \eta \frac{\partial^2 W}{\partial t^2} \tag{2-5}$$

Substituting (2-1) into (2-5), we obtain

$$\frac{\partial^2 W}{\partial z^2} = \frac{\eta}{\nu + 2\mu} \frac{\partial^2 W}{\partial t^2} \tag{2-6}$$

This second-order differential equation is called the wave equation. The solution
for this equation as was discussed has the form of $f(z \pm ct)$ where the negative
sign indicates a wave traveling in the Z-direction whereas the positive sign indicates a wave traveling in the $-Z$-direction. Since the displacement is in the same
direction as wave propagation, this type of wave is called a compressional or longitudinal wave. The sinusoidal solution for this equation is

$$\overset{\pm}{W} = \overset{\pm}{W}_0 e^{j(\omega t \mp kz)} \tag{2-7}$$

where $\omega = 2\pi f$, angular frequency; $k = \omega/c$, the wave number; and $c = \left(\frac{\nu + 2\mu}{\eta}\right)^{1/2}$, phase velocity. For fluid $\mu = 0$, $c = \left(\frac{B}{\eta}\right)^{1/2}$ $\tag{2-8}$

### 2.  Shear Wave

If we assume that in Eq. (2-4) $t_{zz} = t_{zy} = 0$, a new type of wave with the displacement perpendicular to the direction of propagation is characterized by

$$\frac{\partial t_{zx}}{\partial x} = \eta \frac{\partial^2 W}{\partial t^2} \tag{2-9}$$

Since $t_{zx} = \mu(\partial W/\partial x)$, we obtain

$$\frac{\partial^2 W}{\partial x^2} = \frac{\eta}{\mu} \frac{\partial^2 W}{\partial t^2} \tag{2-10}$$

which describes a wave traveling in the X-direction with a displacement in the Z-direction. The sinusoidal solution to Eq. (2-10) is

$$\overset{\pm}{W} = \overset{\pm}{W}_0 e^{j(\omega t \mp k_t x)} \tag{2-11}$$

This type of wave expressed by (2-11) is called shear or transverse wave. The wave number for the shear wave is given by

$$k_t = \frac{\omega}{c_t} \tag{2-12}$$

with propagation velocity given by

$$c_t = \left(\frac{\mu}{\eta}\right)^{1/2} \tag{2-13}$$

It is obvious from Eq. (2-13) that a shear wave can only exist in a medium with nonzero shear modulus; that is, fluid cannot support the propagation of a shear wave.

## C.   Characteristic Impedance

The medium velocity or particle velocity in the Z-direction $u_z$ can be found from the particle displacement $W$ by differentiating $W$ with respect to $t$, that is,

$$u_z = \frac{\partial W}{\partial t} = j\omega W \tag{2-14}$$

for sinusoidal excitation. It can be seen that the particle velocity is always 90° out of phase with respect to the displacement. Since pressure is related to the stress by the following equation,

$$p = -t_{zz}$$

we have

$$p = -(\nu + 2\mu)\frac{\partial W}{\partial z}$$

Since

$$\overset{\pm}{W} = \overset{\pm}{W}_0 e^{j(\omega t \mp kz)}$$

we have

$$p^\pm = \pm jk(\nu + 2\mu)\overset{\pm}{W}$$

or

$$p^\pm = \pm j\omega\eta c\overset{\pm}{W} \tag{2-15}$$

Therefore,

$$p^\pm = \pm \eta c u_z \tag{2-16}$$

Note that the pressure, like the velocity, is 90° out of phase with the displacement. Equation (2-16) also indicates that the pressure and velocity are in phase for the positive-traveling wave, and 180° out of phase for the negative-traveling wave.

The characteristic acoustic impedance of a medium is defined as

$$Z^\pm = \frac{p^\pm}{u_z^\pm} = \pm \eta c \tag{2-17}$$

The acoustic velocity and impedance for a few common materials and biological tissues are listed in Table IV (Goss *et al.*, 1978). The acoustic velocity is a sensitive function of temperature but its dependence on frequency is minimal over the frequency range of diagnostic ultrasound, which is from 1 to 25 MHz.

## D.  Intensity

The intensity of a wave is defined as the average power carried by a wave per unit area normal to the direction of propagation over time. It is well known that power consumed by a force $F$ that has moved an object by a distance $l$ in time $t$ is equal to $Fl/t$. An ultrasound is a pressure wave. Intuitively, one may deduce from the above relationship that the power $P$ carried by an ultrasonic wave is given by

$P$ = force exerted by the pressure wave × medium displacement/time, or
  = force × medium velocity

Now since intensity $i$ is power carried by the wave per unit area, it follows that

$$i(t) = p(t)u(t)$$

For the case of sinusoidal propagation the average intensity can be found by averaging $i(t)$ over a cycle.

$$I = p_0 u_0 \cdot \frac{1}{T} \int_0^T \sin^2 \omega t \, dt = \tfrac{1}{2} p_0 u_0 \quad \text{(2-18)}$$

where $p_0$ and $u_0$ denote peak values of pressure and medium velocity, and $T$ is the period. Similar results can be obtained using complex notations that are

$$p = p_0 e^{j\omega t}, \qquad u = u_0 e^{j\omega t}$$
$$I = \frac{1}{2}[\text{Real part of } pu^*]$$
$$= \tfrac{1}{2} p_0 u_0$$

## Table IV
Velocity and acoustic impedance of pertinent materials and biological tissues at room temperature (20–25°C)

|  | Velocity (m/sec) | Impedance × 10⁻⁶ (kg/m²-sec)[a] |
|---|---|---|
| Water | 1484 | 1.48 |
| Aluminum | 6420 | 17.00 |
| Air | 343 | 0.0004 |
| Plexiglas | 2670 | 3.20 |
| Blood | 1550 | 1.61 |
| Myocardium (perpendicular to fibers) | 1550 | 1.62 |
| Fat | 1450 | 1.38 |
| Liver | 1570 | 1.65 |
| Kidney | 1560 | 1.62 |
| Skull bone | 3360 (longitudinal) | 6.00 |

[a] Rayl is a unit commonly used for acoustic impedance. One rayl = 1 kg/m²-sec.

where * indicates the complex conjugate. Since $Z = p/u = \eta c$, it follows that

$$I = \tfrac{1}{2}\eta c u_0^2 \qquad (2\text{-}19)$$

In terms of particle displacement, we have

$$I = \tfrac{1}{2}\eta c \omega^2 W_0^2 \qquad (2\text{-}20)$$

where $W_0$ is the peak displacement.

Here it is appropriate to define a few terms related to ultrasound intensity that have been used frequently in medical ultrasound as indicators of exposure level. These definitions are necessary because currently a majority of the ultrasonic imaging devices is of the pulse–echo type in which very short pulses of ultrasound consisting of a few cycles of the oscillation are transmitted. This is illustrated in Fig. 67. Therefore, the temporal averaged intensity differs from that given by Eq. (2-18). Moreover, the intensity within an ultrasound beam generally is not spatially uniform. The typical profile of an ultrasonic beam is shown in Fig. 67(c). The spatial average intensity is defined as the average intensity over the ultrasound beam. In Fig. 67(c), it can be seen that the spatial peak intensity in the beam is 1.0 watt/cm² while the spatial average intensity is only 0.3 watt/cm².

Temporal average intensity is defined as the average intensity over a pulse repetition period and is given by the product of duty factor and temporal peak intensity where the duty factor is defined as

$$\text{duty factor} = \frac{\text{pulse duration}}{\text{pulse repetition period}}$$

In Fig. 67, it can be seen that the duty factor is 0.1. So whenever biological effects of ultrasound are considered, it is absolutely necessary to state or understand which definition of intensity is being used (AIUM, 1984). Generally, spatial average temporal average (SATA) intensity is preferred. Sometimes, however, peak values are also used, such as spatial peak temporal average intensity (SPTA).

Althouth these parameters are adequate to provide a rough assessment of the ultrasound exposure level, it has been recommended by the American Institute of Ultrasound (Rockville, Maryland) that in addition to these parameters, peak pressure, spatial peak pulse average intensity [I(SPPA) which is defined as the intensity measured at spatial peak averaged over the pulse duration], total power emitted by the transducer, and other relevant data should also be reported for completeness since the averaged intensity at the focal point of the transducer may be low but the instantaneous intensity or pressure at the focal point and the intensity near the transducer may be high enough to cause bioeffects. As an example, a transducer emitting a total power of 1.1 mW at 3 kHz pulse repetition rate and having a I(SPTA) of 1.1 mW/cm² at the focal point may have a I(SPPA) of 25 W/cm² and peak pressure of 0.66 MPa at the focal point, which are quite high.

## E.  Radiation Force

An acoustic wave exerts a force on any interface across which there is a decrease in ultrasonic intensity in the direction of wave propagation. For a plane wave, the radiation force per unit area or radiation pressure $f_r$ is given by

$$f_r = D(I/c) \qquad (2\text{-}21)$$

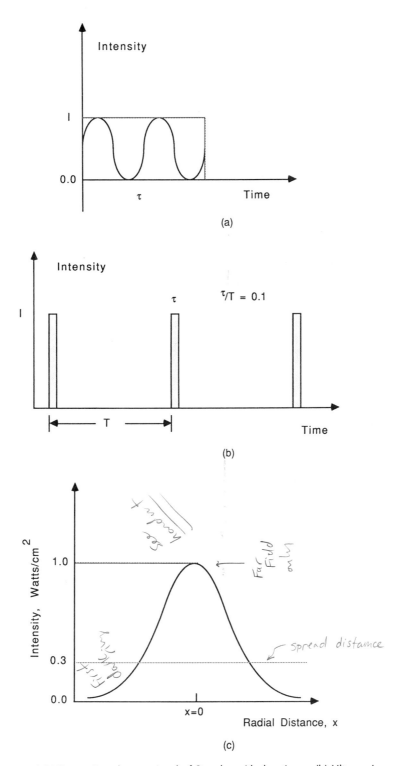

**Figure 67**  (a) Ultrasonic pulse consisted of 2 cycles with duration $\tau$. (b) Ultrasonic pulse train with pulse duration $\tau$ and pulse repetition period $T$. (c) Lateral profile of ultrasound beam. Beam center is at $x = 0$ where intensity $= 1$ watt/cm$^2$.

where $D$ is a constant depending on the physical configuration. If one medium is water and the other medium is one of a very large acoustic impedance such as steel, $D = 2$. On the other hand, if the other medium is a perfect absorber, $D = 1$. The reason for this can be understood by considering the momentum transfer that occurs at the boundary, that is,

$$f_r \, \Delta t = \Delta(\eta u) \tag{2-22}$$

where $\eta$ is the density and $u$ is the particle velocity of water. Since the momentum transfer that occurs at the interface for a perfect reflector is twice as large as that for an absorber, $f_r$ acting on a perfect reflector should be twice as large.

The radiation force is a very useful parameter for determining power of an ultrasonic beam. A device which has often been used for measuring the intensity of an ultrasonic beam is shown in Fig. 68. A target with the surface either covered with an absorbent or a reflective material is suspended in water. The ultrasonic radiator is immersed in water and the beam is normally incident on the target. The weight determined by the balance would yield the radiation pressure, from which the intensity can be calculated from Eq. (2.21).

## F.   Reflection and Refraction

When a wave meets an interface between two different media I and II of acoustic impedance $Z_1$, and $Z_2$, it will be reflected and refracted. The reflected wave returns in the backward direction at the same velocity as the incident wave. The transmitted or refracted wave continues to move in the forward direction but at a different velocity. Just as in optics, the Snell law applies if the wavelength of the wave is much smaller than the dimension of the interface. In Fig. 69, the sub-

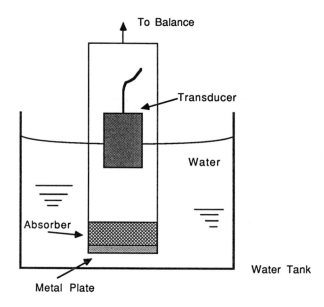

**Figure 68**  Radiation force balance for measuring power contained in ultrasound beam.

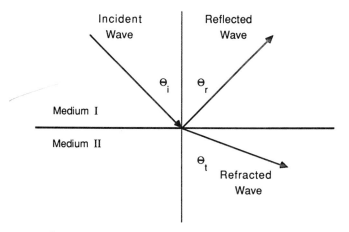

**Figure 69** Reflection and refraction of plane wave at a flat boundary.

scripts i, r, and t refer to incident, reflected, and transmitted or refracted waves, respectively. As in optics,

$$\theta_i = \theta_r \quad \text{and} \quad \sin\theta_i / \sin\theta_t = c_1/c_2 \qquad (2\text{-}23)$$

when $\theta_t = \pi/2$, $\sin\theta_t = 1$, and $\theta_{ic} = \sin^{-1} c_1/c_2$ if $c_2 > c_1$. For any incident angle greater than $\theta_{ic}$, there is no transmission, that is, total reflection occurs. Therefore, $\theta_{ic}$ is called the critical angle.

The pressure reflection and transmission coefficients $R$ and $T$ can easily be found by using the boundary conditions that the pressure and particle velocity should be continuous across the boundary (Brekhovskikh, 1960).

$$R = p_r/p_i = \frac{Z_2 \cos\theta_i - Z_1 \cos\theta_t}{Z_2 \cos\theta_i + Z_1 \cos\theta_t} \qquad (2\text{-}24)$$

$$T = p_t/p_i = \frac{2Z_2 \cos\theta_i}{Z_2 \cos\theta_i + Z_1 \cos\theta_t} \qquad (2\text{-}25)$$

At normal incidence, $\theta_i = \theta_t = 0$

$$R = p_r/p_i = (Z_2 - Z_1)/(Z_2 + Z_1) \qquad (2\text{-}26)$$
$$T = p_t/p_i = 2Z_2/(Z_2 + Z_1) \qquad (2\text{-}27)$$

Since $p = Zu$, $I = p_0^2/2Z$, from Eq. (2-24) and (2-25), it can be shown that

$$I_r/I_i = \left(\frac{Z_2 \cos\theta_i - Z_1 \cos\theta_t}{Z_2 \cos\theta_i + Z_1 \cos\theta_t}\right)^2 \qquad (2\text{-}28)$$

and

$$I_t/I_i = \frac{4Z_2 Z_1 \cos^2\theta_i}{(Z_2 \cos\theta_i + Z_1 \cos\theta_t)^2} \qquad (2\text{-}29)$$

The quantities $(I_r/I_i)$ and $(I_t/I_i)$ are respectively intensity reflection and transmission coefficients of the interface.

At normal incidence, $\theta_i = \theta_t = 0$, we have

$$I_r/I_i = \left(\frac{Z_2 - Z_1}{Z_2 + Z_1}\right)^2 \qquad (2\text{-}30)$$

and

$$I_t/I_i = \frac{4Z_2 Z_1}{(Z_1 + Z_2)^2} \qquad (2\text{-}31)$$

As an acoustic wave propagates through an inhomogeneous medium such as biological tissues, part of its energy will be lost due to absorption and scattering, which will be discussed later, and part of its energy will be lost due to specular reflection at the boundary of two adjacent layers of tissues. The ultrasonic images are formed from the specularly reflected echoes due to planar interfaces as well as the diffusely scattered echoes due to small inhomogeneities in tissue parenchyma. Therefore, any change in the elastic properties of the tissues as a result of a disease may be detectable from their ultrasonic image. This has been the principal rationale behind the conventional ultrasonic imaging techniques.

Since both scattering and reflection depend on the elastic properties of the tissues, which determine the acoustic impedance of the tissue, intuitively one may postulate that echographic visualizability of tissues is determined mostly by their connective tissue content (Fields and Dunn, 1973). The acoustic impedances of connective tissues and tissues containing high concentrations of connective tissues have been found to be much higher than other types of tissue components, for example, fat and protein and tissues containing less connective tissues. The velocity in collagen has been determined to be approximately $1.7 \times 10^5$ cm/sec, while the velocity in elastin is almost unknown (Goss and O'Brien, 1979). The acoustic impedance of blood vessels, which are composed of mainly connective tissues, should be and was found to be higher than most tissues (Geleski and Shung, 1982). Because of their higher impedance than surrounding tissues, blood vessels should be more visualizable on an ultrasonic image than other soft tissues. On the other hand, it has to be noted that the attenuation of ultrasound in connective tissues is quite high as well (Goss and O'Brien, 1980). Very little energy can transmit through a mass such as a solid tumor composed mainly of connective tissues. Thus an acoustic shadow may be created behind the mass. This has been used as one of the criteria for diagnosing tumors.

While in a global sense this postulation that tissue echographic visualizability is largely determined by its connective tissue content may be true, current data suggest that variation of elastic tissue content among major organs such as liver, kidney, spleen, etc., is too small to make any significant difference in attenuation or scattering. Other factors such as cellular dimension and tissue complexity may also play important roles in determining tissue echographic appearance (Fei and Shung, 1985).

There has been a long-standing interest in correlating acoustic parameters such as scattering, attenuation, and acoustic impedance to the biological composition of the tissues or ultrasonic characterization of biological tissues (Greenleaf, 1986).

## G.   Attenuation, Absorption, and Scattering

### 1.   Attenuation

When an ultrasonic wave propagates through a heterogeneous medium, its energy is reduced as a function of distance. The energy may be diverted by reflection or scattering or absorbed by the medium and converted to heat in general. The reflection and the scattering of a wave actually are referring to the same phenomenon, the redistribution of energy from the primary incident direction into other directions. When the wavelength of the wave is much smaller than the object, we call this redistribution phenomenon reflection. On the other hand, if the wavelength is much larger than or comparable to the dimension of the object, we call it scattering.

The pressure of a plane wave propagating in the Z-direction decreases exponentially as a function of $z$ just like X-ray.

$$p(z) = p(0)e^{-\beta z} \tag{2-32}$$

where $p(0)$ is the pressure at $z = 0$ and $\beta$ is the pressure attenuation coefficient. Therefore,

$$\beta = \left(\tfrac{1}{z}\right) \ln\left[\frac{p(0)}{p(z)}\right] \text{ nepers/cm} \tag{2-33}$$

The attenuation coefficient is sometimes expressed in units of dB/cm $\beta'$ or

$$\beta\,(\text{np/cm}) = 20(\log_{10} e)\beta'(\text{dB/cm}) = 8.686\beta'(\text{dB/cm}) \tag{2-34}$$

Typical values of the attenuation coefficient in some materials are given in Table V.

The relative importance of absorption and scattering to attenuation of ultrasound in biological tissues is a matter being continuously debated. Early investigations indicated that scattering contributes approximately 20% of attenuation in liver (Pauly and Schwan, 1970). However, recent results showed that the scattering contribution is considerably less, probably in the order of a few percent (Parker, 1983). Therefore, it is safe to say at this point that absorption is the dominant mechanism for ultrasonic attenuation in biological tissues.

## 2. Absorption

As was discussed earlier, part of energy will be lost due to redistribution of the energy, such as scattering and reflection, and part of the energy will be absorbed by the medium as an acoustic wave propagates through an inhomogeneous medium. The energy absorbed by the medium is generally converted to heat. The absorption mechanisms in biological tissues are quite complex and have been as-

**Table V**
Attenuation coefficients of biological tissues and pertinent materials

| Material | Attenuation coefficient (np/cm at 1 MHz at 20°C) |
|---|---|
| Air | 1.38 |
| Aluminum | 0.0021 |
| Plexiglas | 0.23 |
| Water | 0.00025 |
| Fat | 0.06 |
| Blood | 0.02 |
| Myocardium (perpendicular to fiber) | 0.35 |
| Liver | 0.11 |
| Kidney | 0.09 |
| Skull bone | 1.30 |

sumed to arise from (1) classical absorption due to viscosity and (2) a relaxation phenomenon. Both mechanisms depend on the frequency of the wave. In earlier developments an ideal fluid with $\mu = 0$ has been assumed. Here we will show that this means that absorption due to the classical viscous loss is ignored. However, in reality, this is seldom the case. Recall the definition of shear strain in Fig. 65(b).

$$\epsilon_{xz} = \frac{\partial U}{\partial z} \qquad (2\text{-}35)$$

In studying fluid, we are more interested in the rate of strain rather than strain itself. Hence, differentiating $\epsilon_{xz}$ with respect to $t$, we obtain

$$\frac{\partial \epsilon_{xz}}{\partial t} = \frac{\partial}{\partial z} \frac{\partial U}{\partial t} = \frac{\partial}{\partial z} u_x \qquad (2\text{-}36)$$

where $u_x = \dfrac{\partial U}{\partial t}$ is the particle velocity in the $X$-direction and $\dfrac{\partial}{\partial z} u_x$ is the velocity gradient along the $Z$ axis. When a fluid with finite viscosity is subject to shear stress $t_{xz}$, as shown in Fig. 70, it exhibits a velocity gradient $\partial u_x/\partial z$.

The coefficient of viscosity, denoted as $\zeta$ with units in poises, is defined as the ratio of shear stress to the resultant velocity gradient.

$$\zeta = \frac{t_{xz}}{\partial u_x/\partial z} \qquad (2\text{-}37)$$

It has been shown that for a homogeneous medium like water the absorption coefficient for an ultrasonic wave of frequency $\omega$ is related to viscosity and frequency by the following expression:

$$\beta = \frac{2\omega^2 \zeta}{3\eta c}$$

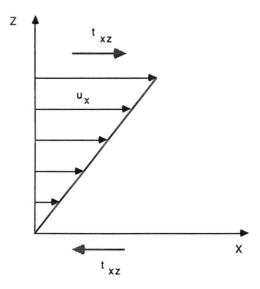

**Figure 70**  Fluid of finite viscosity exhibits a particle velocity gradient when subjected to shear stress.

where $\zeta = \mu/j\omega$ from Eqs. (2-3) and (2-37). Thus, $\beta = 0$ if $\mu = 0$. Note here that the absorption is proportional to $\omega^2$. In many materials, such as air and water, this (frequency)$^2$ dependence of absorption is seen, but in most biological materials, this is not true. The absorption of ultrasound in biological tissues has been hypothesized to be caused by a relaxation process.

**The Relaxation Process**    When a molecule is pushed to a new position by a force and then released, a finite time is required for the molecule to return to its neutral position. This time is called the relaxation time of the molecule. For a medium that is composed of the same type of molecules, the relaxation time is also the relaxation time of the medium. If the relaxation time is short compared to the period of the wave, its effect on the wave will be small. However, if the relaxation time is comparable to the period of the wave, the molecule may not be able to completely return before a second compression arrives. When this occurs, the compressional wave is moving in one direction and the molecules in the opposite direction. More energy is required to reverse the direction of the molecules. On the other hand, if the frequency is increased high enough that the molecules simply cannot follow the wave motion, the relaxation effect again can be neglected. Maximum absorption occurs when the relaxation motion of the particles is completely out of synchronization with the wave motion. Therefore, the characteristics of a relaxation process are that the absorption is maximum at a certain frequency and is negligible in low-frequency and high-frequency regions (Fig. 71). Mathematically, this can be represented by the following equation:

$$\beta_R = \frac{B_0 f^2}{1 + (f/f_R)^2}$$

where $\beta_R$ = absorption coefficient due to the relaxation process, $f_R$ is relaxation frequency $(1/\tau)$, $\tau$ is relaxation time, and $B_0$ is a constant.

Consequently, the absorption coefficient in a tissue can be expressed as

$$\frac{\beta}{f^2} = A + \sum_i \frac{B_i}{1 + (f/f_{Ri})^2}$$

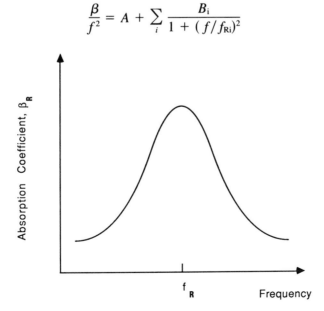

**Figure 71**    Ultrasonic absorption caused by relaxation as a function of frequency.

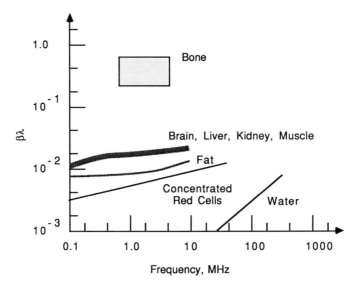

**Figure 72** Product of ultrasonic absorption and wavelength in several media. (From Carstensen, 1979).

where $A$ is a constant associated with classical absorption and $B_i$'s and $f_{Ri}$'s are the relaxation constants and frequencies associated with different species of molecules. Figure 72 shows the absorption of ultrasound in various biological tissues. As can be seen, the absorption is more or less linearly proportional to frequency ($\beta\lambda$ is constant) in diagnostic ultrasound frequency range. A possible explanation for this is illustrated in Fig. 73. Because a tissue has many different types of molecules, many relaxation processes may overlap, giving rise to a flat $\beta\lambda$ in the diagnostic ultrasound frequency range. Also in Fig. 72, it can be seen that the absorption of ultrasound in water is proportional to $f^2$.

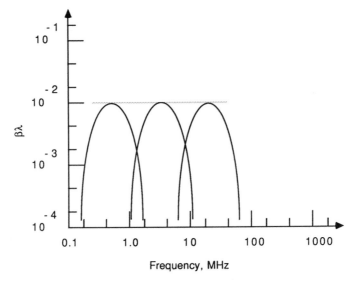

**Figure 73** Superposition of many relaxation responses can produce a flat $\beta\lambda$ curve.

## 3. Scattering

When a wave is incident on an object as shown in Fig. 74, part of the wave will be scattered and part will be absorbed by the object. The scattering characteristics of an object are most conveniently expressed by the term, scattering cross section.

Let us assume the incident pressure is a plane wave $p_i(\mathbf{r}) = e^{j\mathbf{k}\cdot\mathbf{r}}$, where $\mathbf{k}(k = k\mathbf{i})$ and $\mathbf{r}$ are vectors representing the wave number of the incident wave and the position vector. The scattered wave has the following form:

$$p_s(\mathbf{r}) = f(\mathbf{o}, \mathbf{i})\frac{e^{ikR}}{R}p_i(\mathbf{r}_0) \tag{2-38}$$

where $\mathbf{i}$ and $\mathbf{o}$ are unit vectors indicating directions of incidence and observation, provided that the observation point is in the far field of the scatter and $R = |\mathbf{r} - \mathbf{r}_0|$ if $kR \gg 1$.

$f(\mathbf{o}, \mathbf{i})$ in Eq. (2-38) is called the scattering amplitude that describes the scattering properties of the object. It depends on the incident direction and the direction of observation. The incident intensity in a medium of acoustic impedance $z$ is given by

$$I_i = \frac{1}{2}\frac{|p_i|^2}{Z} \tag{2-39}$$

The scattered intensity is given by

$$I_s = \frac{1}{2}\frac{|p_s|^2}{Z} \tag{2-40}$$

Substituting (2-38) and (2-39) into (2-40) and rearranging the equation,

$$I_s = [|f(\mathbf{o}, \mathbf{i})|^2/R^2] \cdot I_i$$

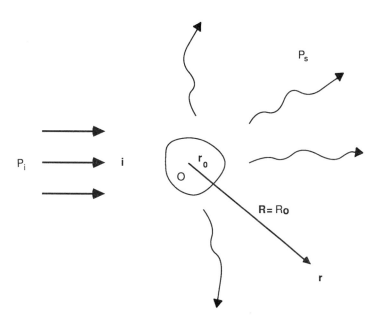

**Figure 74**   Scattering of plane incident wave by scatter O located at $\mathbf{r}_0$.

The differential scattering cross section $\sigma d(\mathbf{i}, \mathbf{o})$, which is defined as the power scattered in $\mathbf{o}$ direction with the incident direction $\mathbf{i}$ per solid angle per unit incident intensity, can be obtained from Eq. (2-40) and is given by

$$\sigma_d(\mathbf{i}, \mathbf{o}) = |f(\mathbf{i}, \mathbf{o})|^2$$

when $\mathbf{o} = -\mathbf{i}$, $\sigma_d(\mathbf{i}, -\mathbf{i})$ is called the backscattering cross section. The scattering cross section $\sigma_s$, defined as the power scattered by the object per unit incident intensity, is then given by

$$\sigma_s = \int_{4\pi} \sigma_d \, d\Omega = \int_{4\pi} |f(\mathbf{i}, \mathbf{o})|^2 \, d\Omega$$

where $d\Omega$ is the differential solid angle.

Similarly, we can define the absorption cross section $\sigma_a$ as the total power absorbed by the object. Then it becomes obvious that the attenuation in wave intensity due to the presence of the object is

$$2\beta = \sigma_a + \sigma_s$$

where $2\beta$ is the intensity attenuation coefficient. This relationship can be obtained in the same way as described in Section I-C in Chapter 1. If there are a number of particles or objects the intensity attenuation coefficient should be

$$2\beta = n \cdot (\sigma_a + \sigma_s)$$

where $n$ is the particle concentration. This relation is valid only if $n$ is small (volume concentration less than 1%). As $n$ increases, multiple scattering occurs and/or there may be particle–particle interactions (Twersky, 1978). This relation is no longer valid under these circumstances.

To solve exactly for the scattering cross section of an object of arbitrary shape is impossible. However, a number of approximations exist that can simplify the problem considerably. One of these is the Born Approximation which assumes that the wave inside the object is the same as the incident wave (Ishimaru, 1978). This is generally a good assumption if the dimension of the object is much smaller than the wavelength, or the acoustic properties of the scatterer are similar to those of the surrounding medium. By applying this approximation and using the wave equation, the scattering cross section of an object can be found. Because of the lengthy nature of the derivation involved, only results will be given here. The scattering cross section for an arbitrary shaped object whose dimension is much smaller than the wavelength is given by (Morse and Ingard, 1968)

$$\sigma_s = \frac{4\pi k^4 a^6}{9} \left[ \left| \frac{G_e - G}{G} \right|^2 + \frac{1}{3} \left| \frac{3\eta_e - 3\eta}{2\eta_e + \eta} \right|^2 \right] \tag{2-41}$$

where $k$ is the wave number, $a$ is the dimension of the particle, $G_e$ and $G$ are the compressibilities of the particle and the surrounding medium, and $\eta_e$ and $\eta$ are the corresponding mass densities.

Equation (2-41) can be applied to calculating the ultrasonic scattering properties of the red blood cells, since the dimension of the red cell ($\sim 3\mu$) is much smaller than the wavelength of ultrasound in the frequency range from 1 to 20 MHz ($\sim 1500\mu$). Using $G_e = 34.1 \times 10^{-12}$ cm²/dyne, $\eta_e = 1.092$ gm/cm³ and those of plasma, which are $G = 40.9 \times 10^{-12}$ cm²/dyne and $\eta = 1.021$ gm/cm³, we find that

$$\sigma_s = 1.1 \times 10^{-12} \text{ cm}^2 \qquad \text{at} \quad 10 \text{ MHz}$$

which is quite small.

The fact that the acoustic scattering characteristics of an object, including angular scattering pattern, depend on the shape, size, and acoustic properties of the scatterer has been known for many years. Ideally, the structure and acoustic properties of the scatterers can be deduced from measuring their scattering properties. This problem is of interest to many scientists in such fields as geophysics, oceanography, and communication. It is generally termed remote sensing or detection.

Although the potential of characterizing tissue structure from its scattering properties was realized in the biomedical ultrasound community almost two decades ago, this field still remains in its infancy primarily due to the complex nature of biological tissues. Preliminary experimental investigations *in vitro* so far have shown that different tissues exhibit different angular scattering pattern and frequency dependence (Linzer, 1976 and 1979). The backscattering coefficient, defined as backscattering cross section per unit volume of scatterers, for five different types of beef tissues as a function of frequency is shown in Fig. 75 (Fei and Shung, 1985). Experimental results on scattering by red blood cells (Yuan and Shung, 1988) are in good agreement with Eq. (2-41). Data on myocardium, on the other hand, show that myocardium has a third power dependence on frequency (Shung and Reid, 1977; O'Donnell *et al.*, 1981). Based on this evidence, it was suggested that the scattering from myocardium may be attributed to myocardial fibers that have a dimension around 10 $\mu$ because the scattering from cylinders whose radius is much smaller than the wavelength is proportional to $f^3$.

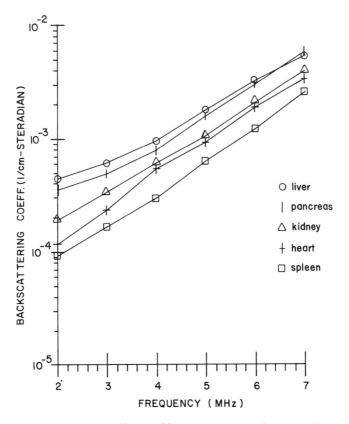

**Figure 75** Backscattering coefficient of bovine tissues as a function of frequency.

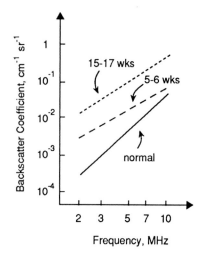

**Figure 76**   Backscattering coefficient of canine myocardium as a function of frequency. Solid line, normal; dashed line, 5–6 weeks after coronary occlusion; dotted line, 15–17 weeks after coronary occlusion. (From O'Donnell *et al.*, 1981).

These results seem to lend some support to the hypothesis that scattering from a tissue may be dependent on the dimension of cellular elements in the tissue (Fei and Shung, 1985).

Since pathological processes in tissues involve anatomical variations, it is likely that they will result in corresponding changes in ultrasonic backscatter. This has been demonstrated by several recent investigations. Figure 76 shows that the backscattering coefficient for regions of infarcted myocardium is substantially higher than that for normal myocardium (O'Donnell *et al.*, 1981). In addition, myocardial ultrasonic backscatter is shown to be related to the contractional state of the tissue (Fig. 77). The myocardial backscatter was found to be highest at end

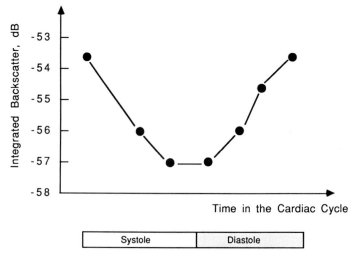

**Figure 77**   Integrated backscatter defined as the averaged backscatter coefficient over a frequency band relative to that from a flat reflector of canine myocardium measured *in vivo* as a function of cardiac cycle. (From Miller *et al.*, 1985).

diastole and lowest at end systole during a cardiac cycle, and this cyclic behavior was blunted in ischemic heart (Miller *et al.*, 1985).

In summary, the attenuation of ultrasound in biological tissues can be attributed to two major mechanisms: (1) scattering and (2) absorption. The question remains, however, as to the relative importance of these mechanisms, although absorption is believed to be dominant. Attenuation generally is not desirable because it limits the depth of penetration of ultrasound into the body. However, it may yield useful information for diagnostic purposes because it carries information about the properties of the tissues if it can be accurately estimated. Work by O'Donnell *et al.* (1979) indicated that attenuation of infarcted myocardium measured *in vitro* is greatly increased over that of normal myocardium for frequencies greater than 5 MHz, and the slope of attenuation is also changed (Fig. 78). Pathology was also found to affect attenuation of other tissues. Unfortunately, measuring attenuation *in vivo* has been proven to be quite difficult a task. Various schemes are being studied, although limited success has been achieved. This topic will be touched on again later in this chapter.

## H.  Nonlinearity Parameter *B/A*

Since the spatial peak/temporal peak intensity I(SPTP) of modern ultrasonic diagnostic instruments can sometimes reach the level of more than 100 W/cm$^2$, nonlinear acoustic phenomena may not be ignored in treating ultrasonic propagation in tissues (Law *et al.*, 1983). Although finite amplitude acoustics has been in existence for a long time, it has not been given any attention in diagnostic ultrasound until recently. The reason for the interest in this phenomenon is twofold: (1) New tissue parameters may be derived for tissue characterization; and (2) nonlinearity can influence, to a significant extent, how the ultrasound energy is absorbed by the tissue.

The nonlinear behavior of a fluid medium can be expressed by a second-order parameter *B/A*. For an adiabatic process in which the entropy is constant or there is no energy flow, the relation between pressure and density can be expressed as a Taylor series expansion of pressure *p* about the point of equilibrium density $\eta_0$ and entropy $s_0$,

$$p = p_{s_0, \eta_0} + A\left(\frac{\eta - \eta_0}{\eta_0}\right) + \frac{1}{2}B\left(\frac{\eta - \eta_0}{\eta_0}\right)^2 + \cdots$$

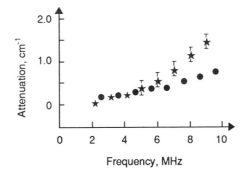

**Figure 78**  Attenuation coefficient of normal (●) and infarcted (⋆) canine myocardium measured *in vitro* as a function of frequency. (From O'Donnell *et al.*, 1979).

where

$$A = \eta_0 \left( \frac{\partial p}{\partial \eta} \right)_{s_0, \eta_0}$$

and

$$B = \eta_0^2 \left( \frac{\partial^2 p}{\partial \eta^2} \right)_{s_0, \eta_0}$$

From this equation and the definition of sound speed,

$$c^2 = \left( \frac{\partial p}{\partial \eta} \right)_{s_0, \eta_0}$$

we can show that $B/A$ is given by

$$B/A = 2\eta_0 c_0 \left( \frac{\partial c}{\partial p} \right)_{s_0, \eta_0} \tag{2-42}$$

Equation (2-42) can be converted to parameters that are easier to measure using thermodynamic relationships.

$$B/A = 2\eta_0 c_0 \left( \frac{\partial c}{\partial p} \right)_{T, s} + 2 \frac{c_0 T \beta_v}{C_p} \left( \frac{\partial c}{\partial T} \right)_{p, s} \tag{2-43}$$

where $T$ is temperature, $C_p$ is the heat capacity per unit mass at constant pressure, and $\beta_v$ is the volume coefficient of thermal expansion. The first term represents the change in sound speed per unit change in pressure at constant temperature and entropy. The second term represents the change in sound speed per unit change in temperature at constant pressure and entropy. Therefore, $B/A$ can be estimated from these quantities, which are either known or measurable. This approach is called the thermodynamic method. It can also be estimated by a finite amplitude method (Cobb, 1983) in which the second pressure harmonic is measured and extrapolated back to the source to eliminate the effect of absorption. The preliminary findings are (1) $B/A$ is linearly proportional to the solute concentration in aqueous solutions of proteins, (2) $B/A$ is insensitive to the molecular weight of the solute at fixed concentrations, (3) $B/A$ ranges from 6 to 11 for soft tissues, and (4) $B/A$ may be dependent on tissue structure.

Since the thermodynamic method is unsuitable for *in vivo* investigation, *in vivo* measurement of $B/A$ is likely to be pursued using the finite amplitude technique.

## I.   Doppler Effect

The Doppler effect describes a phenomenon in which a change in the frequency of sound emitted from a source is perceived by an observer when the source or the observer is moving or both are moving. The reason for the perceived frequency change for a moving source and a stationary observer is illustrated in Fig. 79. In diagram (a), the sound source $S_p$ is stationary and producing a uniform spherical wave, and the frequency of the sound perceived by an observer is

$$f = \frac{c}{\lambda}$$

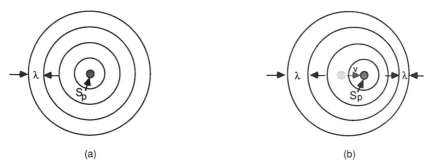

**Figure 79** Doppler effect: (a) stationary source and receiver, (b) moving source and stationary receiver.

where $c$ is the velocity of sound in the medium and $\lambda$ is the wavelength shown as the distance between two crests in the Fig. 79(a). In diagram (b), the sound source is moving to the right with a velocity $v$. It emits a wave peak and then runs after the peak in one direction and away in the opposite direction. Therefore, the source motion changes the distance between the peaks, increasing frequency and decreasing the wavelength to the right and decreasing the frequency and increasing the wavelength to the left. So the frequency perceived by an observer on the right is

$$f' = \frac{c}{\lambda'} = \frac{c}{\lambda - vT} = \frac{c}{(c - v)T} = \frac{c}{c - v}f$$

and the frequency seen by an observer on the left is

$$f' = \frac{c}{c + v}f$$

The difference between the actual frequency of the source $f$ and the perceived frequency $f'$ is called the Doppler frequency, $f_d$. Similar relationships can also be obtained for a moving observer. When these relationships are combined, for a source moving with a velocity $v$ and an observer with velocity $v'$, the Doppler frequency is found to be given by

$$f_d = f' - f = [(c + v')/(c - v) - 1]f \qquad (2\text{-}44)$$

If the source and observer are moving at the same velocity $v$, then Eq. (2-44) can be reduced to

$$f_d = 2vf/c \qquad (2\text{-}45)$$

For the situation where the velocity is making an angle of $\theta$ relative to the direction of sound propagation, $v$ in Eq. (2-45) should be replaced by $v \cos \theta$. The Doppler effect is used in many instrumentation applications and several are important in biomedical instrumentation. For example, the measurements of the blood flow velocity transcutaneously, that is, without penetrating the skin in any manner, can be accomplished by sending ultrasonic waves into the blood vessel, detecting the scattered radiation from the moving red cells, and developing appropriate instrumentation to extract the Doppler frequency, which is proportional to the red cell velocity.

# II.    Generation and Detection of Ultrasound

## A.    Piezoelectric Effect

Certain materials have the property that the application of an electrical field causes a change in their physical dimensions and vice versa. This phenomenon is known as the piezoelectric effect (pressure–electric effect). Certain crystals such as quartz and tourmaline that occur in nature are piezoelectric. The physical reason for the piezoelectric phenomenon can be explained as follows: The piezoelectric materials can be thought as being made up of innumerable electric dipoles, as shown in Fig. 80. An externally applied electrical voltage $V$ will realign the dipoles in a slab of piezoelectric material and thus change the thickness of the slab from $L$ to $L + \Delta L$. The illustration shows a considerable change in the dimension, but in reality the change is only a few microns. Conversely, the application of a stress produces a voltage across the slab surfaces.

Another group of artificial materials, known as polarized ferroelectrics, possess strong piezoelectric properties. Polarization of a ferroelectric material is carried out by heating it to a temperature just above a certain level, depending on the material, known as the Curie temperature and then allowing it to cool slowly in the presence of a strong electric field, typically in the order of 20 kV/cm, applied in the direction in which the piezoelectric effect is required. This process will align the dipoles along the direction of polarization. There is a great variety of ferroelectric materials. Barium titanate was the first to be discovered. It has now been largely replaced by lead zirconate titanate (PZT). Several types of PZT are commercially available. The electrical field is usually applied to the material by means of two silver electrodes, as illustrated in Fig. 81.

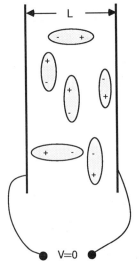
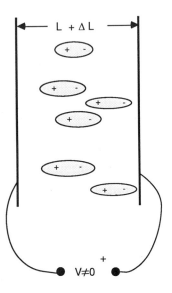
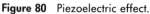

**Figure 80**   Piezoelectric effect.

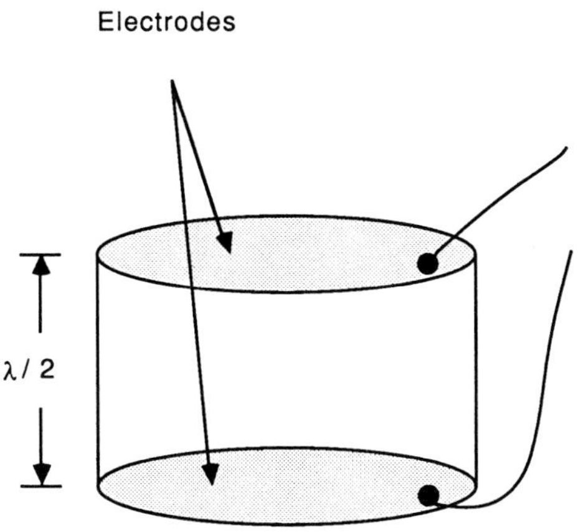

Electrodes

**Figure 81** Piezoelectric disc coated with silver electrodes. Resonance of disc occurs when disc thickness = $\lambda/2$ where $\lambda$ is ultrasound wavelength in disc material.

The stress produced in a piezoelectric material by the application of a unit electric field without strain is called the piezoelectric stress constant $e$ in units of newtons/V-m or $C/m^2$. The transmitting constant or piezoelectric strain constant $d$, in coulombs/newton, is defined as the strain produced per unit applied electric field when external stress = 0 and is related to $e$ by $e = c^E \cdot d$ where $c^E$ is the material's elastic constant under no electric field [see Eq. (2-1)]. On the other hand, the receiving constant $g$ in units of V-m/newton, is the electric field generated under open-circuit condition per unit applied stress. The dielectric constant $\epsilon$ of a piezoelectric material depends on the extent of freedom of the material. Two values are often used in the literature. If the material is so clamped that it cannot move in response to an applied field, the dielectric constant measured is designated as $\epsilon^S$. If the transducer is free to move without restriction, the dielectric constant measured is denoted as $\epsilon^T$. The transmitting constant and the receiving constant are related by the following relationship (Kino, 1987):

$$d = g\epsilon^T$$

The ability of a transducer to convert one form of energy from another is measured by its electromechanical coupling coefficient, defined as

$$\text{electromechanical coupling coefficient} = \frac{\text{stored mechanical energy}}{\text{total stored energy}}$$

It should be noted that this quantity is not the efficiency of the transducer. If the transducer is lossless, its efficiency is 100% but the ECC is not necessarily 100% because some of the energy may be stored as mechanical energy while the rest of the energy may be stored dielectrically in a form of potential energy. The electromechanical coupling coefficient is related to $e$ by the following equation:

$$\text{ECC} = e^2/c^E\epsilon^S = \epsilon^T/\epsilon^S - 1$$

Since only the stored mechanical energy is useful, ECC is a measure of the performance of a material as a transducer.

Here it should be noted that most materials or crystals are anisotropic. There-fore, to completely describe the piezoelectric properties of a material, 18 piezo-electric stress constants and 18 receiving constants are required. Fortunately, since these materials usually are symmetric, a smaller number of constants are actually needed. For instance, in quartz there are only five constants. Depending on the symmetry of the crystal, the principal piezoelectric axes of the crystal $(X, Y, Z)$ can be established. The principal axis is defined as the direction in which stress produces charge polarization parallel to the strain (Bolt and Heuter, 1955). A plate cut with its surfaces perpendicular to the $X$ axis is called $X$-cut, and so forth.

The transmitting and receiving constants for a few materials are listed in Table VI. However, it should be cautioned that this table is not intended to be an exhaustive listing of all pertinent piezoelectric materials. There are several newer ceramic piezoelectrical materials, such as lead metaniobate, which are superior to PZT in certain aspects of the transduction process.

Several recent developments in the field of transducer technology are worth mentioning. One is the development of piezoelectric polymer (Sessler, 1981). Thin film made of the polymer polyvinylidence fluoride ($PVF_2$) has been shown to exhibit piezoelectric properties. $PVF_2$ is a semicrystalline material. After pro-cesses like polymerization, stretching and poling, thin sheets of $PVF_2$ with a thickness in the order of 6 to 50 $\mu$m can be used as transducer material. It is ex-tremely wideband. This material is attractive for medical imaging because its acoustic impedance matches the impedance of the tissue and because it is flexible. In addition, it is inexpensive. The disadvantages are that the dielectric loss is large and the dielectric constant is low. Despite these drawbacks, $PVF_2$ films have found many applications in diagnostic ultrasound (Shung, 1987). Miniature $PVF_2$ hydrophones are commercially available.

A second interesting development involves phase-insensitive ultrasonic trans-ducers for intensity measurement. Phase cancellation effect may occur at the face of a phase-sensitive transducer, such as a ceramic or a quartz transducer, resulting in appreciable errors in the measurement of attenuation and scattering (Marcus and Carstensen, 1975; Busse and Miller, 1981). This phase cancellation effect can be eliminated by using an intensity-sensitive ultrasonic transducer based on the acoustoelectric effect. It has been shown that the acoustic energy can be coupled to the charge carriers in a piezoelectric semiconductor, such as cadmium sulfide.

## Table VI
Piezoelectric properties of several important piezoelectric materials

| Property | $PVF_2$ | Quartz ($X$-cut) | PZT-4 | PZT-5A |
|---|---|---|---|---|
| $d \times 10^{12}$(C/N) | 15 | 2.31 | 289 | 374 |
| $g \times 10^2$ (V/m/N) | 14 | 5.78 | 2.61 | 2.28 |
| $e$(C/m$^2$) | 0.07 | 0.17 | 15.1 | 15.8 |
| ECC | 0.12 | 0.1 | 0.7 | 0.71 |
| $\epsilon^T \times 10^{11}$ (F/m) | 9.7 | 3.98 | 1150 | 1504 |
| $c$(m/sec) | 2070 | 5740 | 4000 | 3780 |
| $\eta$ (kg/m$^3$) | 1760 | 2650 | 7500 | 7750 |
| Curie temp (°C) | 100 | 573 | 328 | 365 |

CdS is a photosensitive semiconductor; conduction electrons can be generated by exposing it to light. These electrons will then follow the motion generated by the acoustic wave. A dc voltage can thus be generated across the crystal. This voltage has been found to be related to the concentration of the conduction electrons and the intensity of the acoustic wave. A prototype acoustoelectric transducer was built and shown to be effective in eliminating phase cancellation (Heyman, 1978).

One of the most promising frontiers in transducer technology is the development of piezoelectric composite materials (Smith, 1989). PZT polymer composites in the form of small PZT rods embedded in low density polymer, which have been under development for low-frequency applications for many years, are now being studied for medical ultrasonic applications. These composites, typically in volume concentration of 20 to 40% PZT, have a lower acoustic impedance (less than 10 Mrayl) than conventional PZT material (greater than 20 Mrayl). The lower acoustic impedance better matches the acoustic impedance of human skin. The composite material can be made flexible and has an electromechanical coupling coefficient comparable to that of PZT (between 0.6 and 0.75). The mechanical and dielectric losses are also lower. Higher coupling coefficient and better impedance matching may lead to higher sensitivity and improved resolution. Recent findings (Gururaja *et al.*, 1985) show that the performance of a composite is directly related to the ratio between the wavelength of the transverse waves produced as a result of the rod vibration in the radial mode and the gap among the rods, regardless of the PZT volume fraction. This means that the performance of composite transducers can be controlled by selecting proper embedding polymer and rod periodicity.

One of the problems associated with composite materials is the higher fabrication cost. Typical procedure involves first dicing PZT and subsequently filling the gaps with polymer. The performances of a number of prototype composite annular arrays with frequencies from 3 to 7.5 MHz have been evaluated and found to be superior to similar PZT devices.

## B. Ultrasonic Transducers

A number of factors are involved in choosing a proper piezoelectric material for transmitting the ultrasonic wave, such as stability, piezoelectric properties, and the strength of the material. Quartz is suitable for accurate measurements because of its stability, but it requires relatively high electric field strength to obtain high power outputs. On the other hand, the ceramics require much lower electric field to produce similar power outputs. They have the disadvantages of lesser stability and low electric input impedance at high frequencies. A typical single-element transducer is shown in Fig. 82. The crystal is located near the face of the transducer. The surfaces of the crystal are plated with silver or gold. The outside electrode is usually grounded to protect the patients from electrical shock. The housing can be metallic or plastic. Acoustic insulator can be placed between the crystal and the housing to prevent ringing of the housing that follows the vibration of the crystal.

By considering the two surfaces of the piezoelectric crystal as two independent vibrators, one can easily see that the resonant frequencies for such a transducer are

$$f_0 = \frac{nc_p}{2L_c} \quad \text{or} \quad L_c = n\lambda/2$$

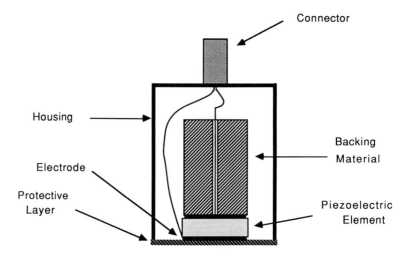

**Figure 82**  Typical construction of single-element transducer.

with the lowest resonant frequency being $n = 1$ where $c_p$ is the acoustic wave velocity in the transducer material, $L_c$ is the thickness of the piezoelectric material, and $n$ is an odd integer ($n = 1, 3, 5, \ldots$).

The transducer can be treated as a three-port network as shown in Fig. 83, two being mechanical ports representing the front and back surfaces of the piezoelectric crystal and one being an electrical port representing the electrical characteristics of the crystal to the electrical generator (Kino, 1987). Various sophisticated equivalent circuits exist to model the behavior of the transducer. The most well known are the Mason model, the Redwood Model, and the KLM model (Kino, 1987). These models can be simplified considerably if some assumptions are made. The equivalent circuit for a transmitting transducer irradiating into water with air backing can be represented by a two-port network consisting of a capacitor and a resistor (Schwan, 1969). The reason for this is that the reflection at the interface of the piezoelectric crystal and air is extremely large so that the acoustic port representing the back surface can be ignored or considered an open circuit. This network is shown in Fig. 84 where $R_r$ is the radiation resistance given by $R_r = L_c^2 Z_w / 4e^2 A_c$ and $C_t$ is the capacitance of the crystal given by $C_t = \epsilon^T A_c / L_c$ where $A_c$, $L_c$, and $Z_w$ are respectively the surface area of the crystal, thickness of the crystal, and the acoustic impedance of water. $\epsilon^T$ and $e$ are

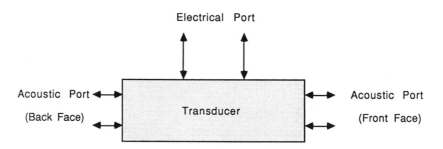

**Figure 83**  Three-port network model of single-element transducer.

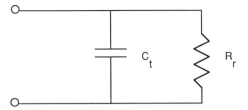

**Figure 84**  Equivalent circuit for air-backed single-element transducer at resonance irradiating into water.

piezoelectric properties of the crystal defined in Section II-A. The average intensity irradiated into water by the transducer driven by a sinusoidal signal at the lowest resonant frequency of the transducer is given by

$$I = \frac{2e^2 V_p^2}{L_c^2 Z_w} \text{ watts/cm}^2 \qquad (2\text{-}46)$$

where $V_p$ is the peak voltage across the faces of the transducer.

Equation (2-46) applies only if the operating frequency is the resonant frequency of the crystal. If the frequency deviates from the resonant frequency for a constant voltage, the irradiated intensity is decreased, according to the $Q$ factor of the resonant circuit. The $Q$ factor of a resonant circuit is defined as

$$Q = f_0/(f_2 - f_1)$$

where $f_0$ is the resonant frequency and $f_2$ and $f_1$ are the frequencies at which the amplitude drops by $-3$ dB relative to the maximum or 3 dB down points (see Fig. 85). Two $Q$'s have been used to describe ultrasonic transducers, namely, electrical $Q$ and mechanical $Q$, since they are electromechanical devices. In ap-

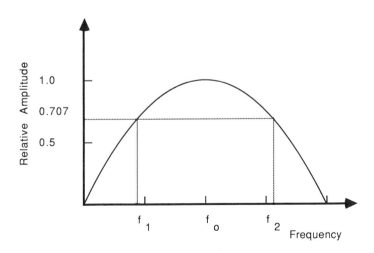

**Figure 85**  Frequency response of a transducer from which transducer $Q$ can be defined.

plications where maximal output is needed and bandwidth requirement is not crucial, as in a continuous-wave ultrasonic Doppler flowmeter, the $Q$ of the transducer should be large whereas in pulse–echo devices, the $Q$ should be small. To adjust the $Q$ or bandwidth of a transducer, either electrical matching, which optimizes electrical $Q$, or mechanical matching, which optimizes mechanical $Q$, can be used.

## 1. Mechanical Matching

When a transducer is excited by an electrical source, it rings according to its resonant frequency. For continuous-wave application, the transducers are air-backed, allowing as much energy irradiated into the forward direction as possible. Due to the mismatch in acoustic impedance between the air and the piezoelectric material, acoustic energy at this interface is reflected into the forward direction. Thus very little energy is lost. On the other hand, this mismatch, which produces the so-called ringing effect for pulse–echo applications, is very undesirable because it lengthens the pulse duration.

Strong and absorbing backing materials can be used to damp out the ringing or to increase bandwidth (to lower the $Q$). The backing material not only absorbs part of the energy from the vibration of the back face but also minimizes the mismatch in acoustic impedance. In general the best backing material is the one that has an acoustic impedance similar to that of the transducer and absorbs as much as possible the energy that enters it. It must be noted that the suppression of ringing or the shortening of pulse duration is achieved by sacrificing sensitivity, because a large portion of the energy is absorbed by the backing material. Various types of backing materials have been used, including a mixture of rubber and tungsten powders in epoxy resin and unpoled piezoelectric materials.

Acoustic insulators can be placed between the case of the probe and the crystal and the backing block assembly to minimize the coupling of the ultrasonic energy into the case. This is sometimes necessary because the case is often made of a low-loss material, such as metal, and is likely to ring in response to an ultrasonic transient.

Figure 86 shows the acoustic pulses generated by the application of a fast electrical transient to transducers (a) with no backing and (b) with heavy backing. The spectrum of an ultrasonic pulse produced by a 5 MHz broadband of a low $Q$ transducer is shown in Fig. 87. The spectrum of an ultrasonic pulse varies as it penetrates into tissue because the attenuation is frequency dependent. It is known that the center frequency and bandwidth decrease as the ultrasound pulse penetrates deeper. In other words, the resolution along the beam, which is proportional to the pulse duration, worsens as the beam penetrates deeper into the tissue. In commercial scanners, pulse shape and duration are maintained by time gain compensation and some form of signal processing, which will be discussed in Section III-A.

It has been shown that the bandwidth of a transducer can be increased by the use of a matching layer of suitable thickness and characteristic impedance between the crystal and mechanical load. Such a layer not only facilitates better transmission of the wave from the transducer to the load, but also broadens the bandwidth. If poor sensitivity and some degree of ripple can be tolerated in the pass band, a transducer matched both to the load and backing has the widest bandwidth.

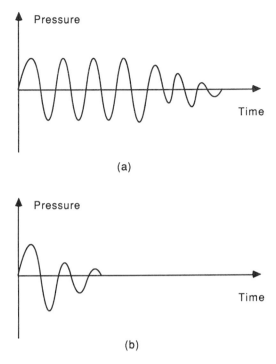

(a)

(b)

**Figure 86** Acoustic pulses produced by transducer (a) with no backing and (b) heavy backing.

## 2. Electrical Matching

Maximizing energy transmission and bandwidth can also be achieved by matching the electrical characteristics of the transducer to the electrical source and amplifier. Circuit components may be placed between the transducer and external electrical devices (DeSilets *et al.*, 1978).

## C. Transducer Beam Characteristics

The field characteristics of an ultrasonic transducer beam are non-ideal. It deviates significantly in many ways from an ideal parallel beam with uniform intensity profile. The beam profile can be calculated utilizing Huygens' principle.

## 1. Huygens' Principle

Huygens' principle states that the resultant wavefront generated by a source of finite aperture can be obtained by considering the source to be composed of an infinite number of point sources. To calculate the beam profile of an ultrasonic transducer, we can approach the problem in similar fashion, that is, by considering the transducer surface to be made of an infinite number of point sources while a spherical wave is emitted by each point source. The superposition of the spherical wavelets generated by all of the point sources on the tranducer surface at a certain point yields the field at that point (Kinsler and Frey, 1962).

## 5 MHz Transducer Pulse

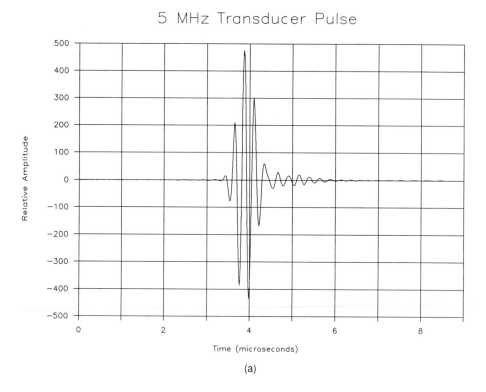

(a)

## 5 MHz Transducer Pulse Spectrum

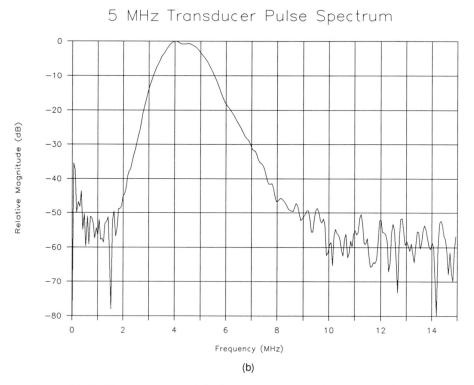

(b)

**Figure 87** (a) Time response and (b) frequency response of a commercial 5-MHz single-element transducer.

## 2.  Beam Profiles

Figure 88 shows the typical axial pressure profile of an ultrasonic transducer driven by a sinusoidal source. It is a plot of the ultrasonic peak pressure as a function of the axial distance from the center of the transducer. The distance between the transducer and the last maximum of the axial pressure is found to be

$$z_0 = \frac{r^2}{\lambda} \tag{2-47}$$

where $r$ is the radius of the element and $\lambda$ is the wavelength. The region where axial pressure oscillates, $z < z_0$, is called the near-field zone or Fresnel zone and the region where the axial pressure decreases approximately according to $1/z$ is called far-field zone, or Fraunhofer zone. Obviously, $z_0$ is the transition point from near-field zone to far-field zone (Wells, 1977).

Figure 67(c) shows the transverse pressure profile of a transducer in the far-field zone. Usually in the far-field, the distance between two points where the pressure drops to $-3$ dB of the maximal value, which is in general the axial pressure, is defined as beam width. The transverse profile in the near field is extremely complex and not well defined. As $z$ becomes greater than $z_0$, the beam starts to diverge, which is illustrated in Fig. 89. The angle of divergence can be calculated approximately using the following formula:

$$\sin \theta = 0.61 \frac{\lambda}{r} \tag{2-48}$$

Figure 90 shows the angular radiation pattern of intensity in the far field of an ultrasonic transducer. The radiation pattern generated by an ultrasonic transducer consists of a main lobe and several side lobes. The first zero or the angle at which the main lobe becomes zero occurs at $\theta = \sin^{-1}(0.61\lambda/r)$. The number of side lobes and their magnitude relative to that of the main lobe depends on the ratio of the wavelength to transducer radius. As the ratio becomes smaller, $\theta$ decreases or the beam becomes sharper accompanied by an increase in the number of side lobes. Side lobes are very undesirable in ultrasonic imaging because they produce spurious signals resulting in artifacts in the image. Therefore, to have a sharper beam by reducing $\lambda/r$, more side lobes are introduced and $z_0$ is shifted farther away from the transducer. Consequently, for a particular application, a compromise has to be reached.

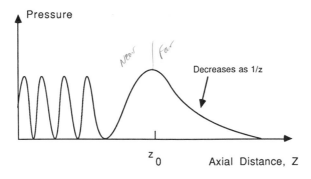

**Figure 88**  Axial peak pressure profile of a single-element transducer.

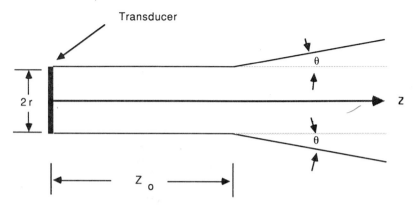

**Figure 89**   Idealistic view of lateral peak pressure profile of a single-element transducer.

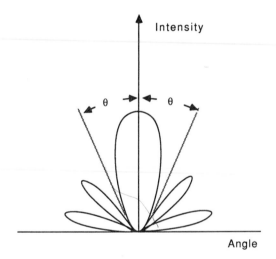

**Figure 90**   Intensity radiation pattern of a single-element transducer.

## 3.   Pulsed Ultrasonic Field

The above discussion pertains only to continuous-wave propagation. Most applications of ultrasound in medicine, however, involve pulsed ultrasound. By using the Fourier transform and the principle of superposition, the field characteristics of a transducer transmitting pulses can be calculated (Rose and Goldberg, 1981). The differences lie primarily in that abrupt variations are moderated. A pulsed ultrasonic axial beam profile is shown in Fig. 91.

## 4.   Visualization and Mapping of the Ultrasonic Field

A schlieren system is an optical system that can be used to visualize the ultrasonic field. This method depends on the diffraction of a ray of light from its undisturbed

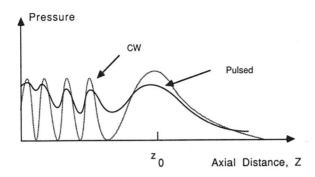

**Figure 91**  Axial beam profiles of a single-element transducer excited by short pulses (solid line) and continuous wave (hashed line).

path when it passes through a medium in which there is a component of refractive index gradient normal to the light beam. This technique is extremely sensitive. A change in refractive index of about 1 part in $10^6$ can be observed. The propagation of an ultrasonic wave is associated with changes in the density of the medium through which it travels. It has been demonstrated that a transparent medium supporting an ultrasonic wave diffracts light. This diffraction occurs because the density of the medium varies in sympathy with the ultrasonic wave, and the refractive index of the medium depends on its density. A typical schlieren system is shown in Fig. 92(a). A beam of light is arranged to pass through a transparent medium, usually water, in which is established the ultrasonic field to be investigated. The light is then focused on an obstruction, so that none reaches the observer when the optical field is not disturbed. However, when ultrasound changes the index of refraction of the medium, the light that passes through the disturbed areas no longer falls at the original focus, but it is deviated so that portion of the light which reaches the observer is changed. The schlieren image may be observed directly on a screen or photographed. Figure 92(b) is a schlieren image of a pulsed ultrasonic beam generated by a transducer.

Qualitative interpretation of the schlieren image is quite simple: The brighter the image appears, the more intense is the corresponding part of the ultrasonic field. Unfortunately, quantitative analysis of the optical image is not as simple. Therefore, to quantitatively map the ultrasound field, nondirectional microprobes or hydrophones may be used. In addition to nondirectional, this type of probe should possess nonselective frequency characteristics or a very broad frequency bandwidth. It should be small in size to avoid the establishment of standing waves. However, in practice, it is almost impossible to satisfy all these requirements. A typical microprobe is shown in Fig. 92(c). The diameter of the probe is less than 1 mm and the transducer is housed at the tip of a hypodermic needle. Both PZT and PVDF polymer have been used as the piezoelectric elements in commercial needle-type hydrophones (Lewin, 1981). Another form, the poled membrane hydrophone, in which only a small spot in the center of a large PVDF membrane is poled to act as ultrasound receiver, has also been developed (McDicken, 1991). Alternatively, a small target such as a small sphere or wire may be used to map the field.

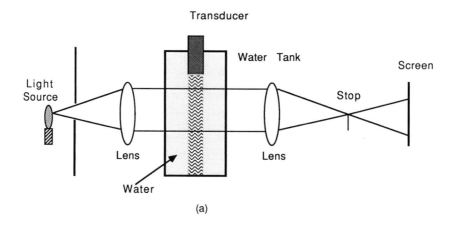

(a)

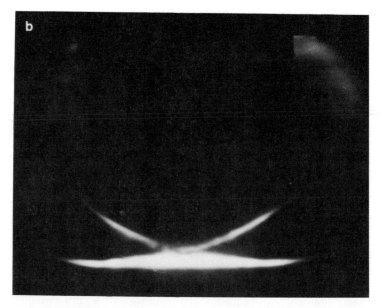

**Figure 92**    (a) Schlieren system for visualizing ultrasonic beam. (b) Schlieren image of rectangular transducer at the top of the diagram excited by 0.5-MHz one-cycle pulse. (Courtesy of John Weight, City University, London, UK) Bright line at the bottom indicates ultrasound pulse whereas two bright circular arcs indicate ultrasound produced by transducer edge. (c) Ultrasonic hydrophone.

c

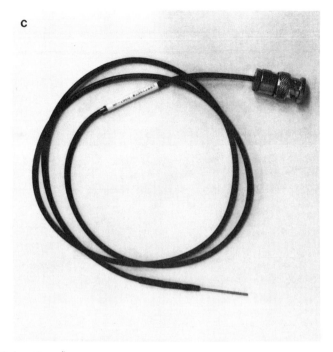

**Figure 92** (*continued*)

## D.   Axial and Lateral Resolution

A transverse pressure profile can be obtained as a transducer scans a small test object, as shown in Fig. 93(a). The lateral resolution of a transducer is determined primarily by the beam width of the transducer ($-3$ dB beam width or $-6$ dB beam width). Why the beam width determines the lateral resolution is illustrated in Fig. 93(b). Transducer A has a narrower beam width, thus a better resolution, than transducer B. If two test objects are $d_1$ apart, both transducers can separate them. If they are $d_2$ apart, transducer A can still separate the two objects but transducer B can only barely distinguish the two. If they are $d_3$ apart, transducer A still can separate the two objects but transducer B cannot.

The actual ultrasound images of the AIUM (American Institute of Ultrasound in Medicine) test object obtained by two transducers with different lateral resolution are shown in Fig. 94(a) and (b). The AIUM test object consisting of a series of fine needles embedded in a medium mimicking tissue ultrasonic pattern is shown in Fig. 94(c). Figure 94(a) shows the image obtained by a focused transducer, whereas Fig. 94(b) shows that for a nonfocused transducer at the same frequency with the same radius. The improved lateral resolution in the focal region is clearly seen.

More sophisticated phantoms have been developed to mimic not only tissue ultrasonic pattern or texture but also its ultrasonic properties including speed, attenuation, and backscatter so that the performance of ultrasonic imaging devices can be better assessed (Madsen, 1986). There are now Doppler phantoms commercially available to evaluate the performance of Doppler devices as well. These

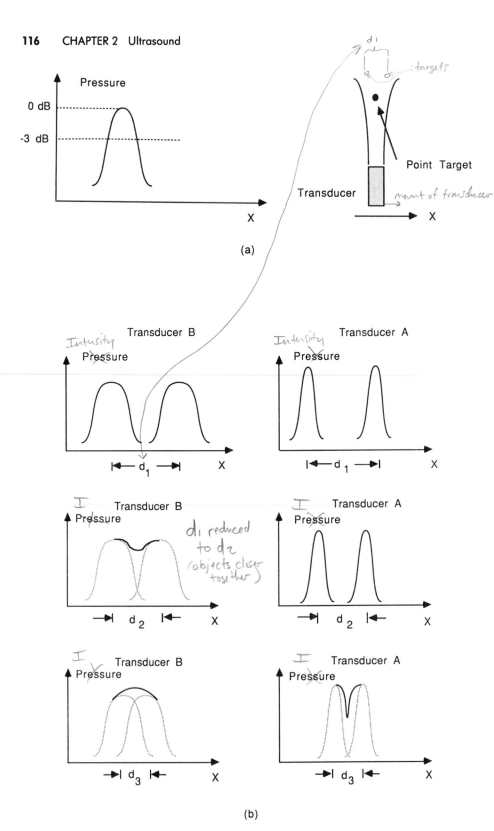

**Figure 93**  (a) Lateral beam profile of transducer can be obtained by scanning transducer across point targets (b) Diagram illustrates why transducer with narrower beamwidth has better lateral resolution.

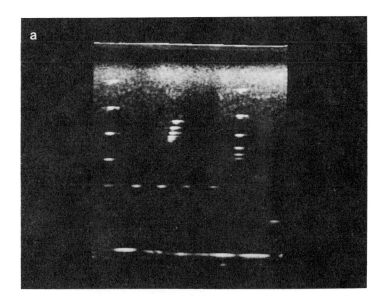

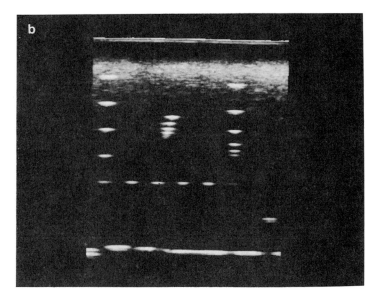

**Figure 94**   B-mode images of AIUM test object obtained with (a) focused
transducer and (b) nonfocused transducer of same diameter. Top and bottom
bright lines indicate echoes from front and back face of phantom housing whereas
shorter lines indicate echoes from wire targets. The widths of these lines are
smaller in middle of image (a) where beam is focused in comparison to image (b).
(c) AIUM test phantom.

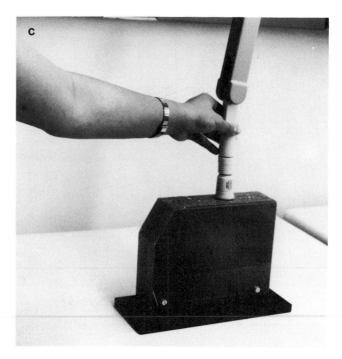

**Figure 94** (continued)

phantoms have become indispensable tools for quality assurance in an ultrasonic imaging laboratory.

The axial or range resolution of a transducer is determined by the spatial extent of the transmitted ultrasonic pulse given by $\tau c$ where $\tau$ is the pulse width. Using the same argument for lateral resolution, it is apparent that axial resolution is related to the pulse width. Since the pulse shape transmitted by a transducer is not well defined (Fig. 86), again the duration between the times where the amplitude of the pulse drops to $-3$ dB of its maximal value may be defined as the pulse width.

## E. Focusing

The primary function of focusing is to improve the lateral resolution, because in certain axial range the beam width is reduced by focusing. However, an improvement in the lateral resolution in a certain range always accompanies a loss of resolution in other regions, as shown in Fig. 95(a).

The general principles of focusing are those that apply in elementary optics. Two most often used schemes are illustrated in Fig. 95. In (b), a self-focusing radiator is shown. In (c), an acoustic lens is shown. Acoustic lenses are usually made of Plexiglas or polystyrene. The focal length $f$ is given by

$$f \approx \frac{\delta}{1 - 1/n} \qquad (2\text{-}49)$$

where $\delta$ is the radius of curvature and $n = c_1/c_2$, $c_1$ being the velocity in the lens and $c_2$, that in the medium. Equation (2-49) is valid only if $\phi$ is small and

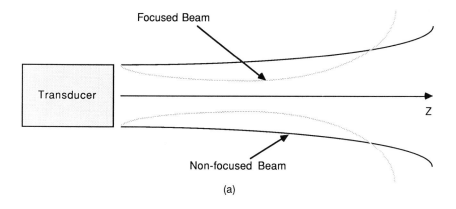

Focused Beam

Transducer

Z

Non-focused Beam

(a)

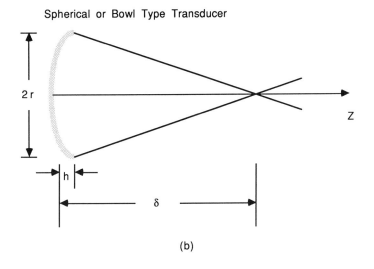

Spherical or Bowl Type Transducer

2 r

Z

h

δ

(b)

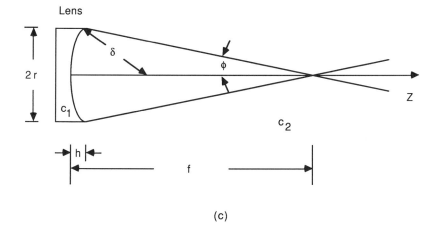

Lens

δ

φ

2 r

Z

$c_1$

$c_2$

h

f

(c)

**Figure 95**   (a) Focusing of transducer reduces beamwidth in limited region of beam but results in more rapid divergence beyond focal zone. (b) Self-focusing transducer. (c) Focusing with aid of lens.

$h < 0.1\delta$ (Wells, 1977). The focal region formed by an acoustic lens is generally ellipsoidal. Its dimension depends on the relationship between wavelength and the diameter of the lens. Generally, the bigger the diameter, the smaller the focal point.

## F.  Arrays

Transducer arrays are novel transducer assemblies with more than one transducer element. These elements may be rectangular in shape and arranged in a line (linear array, Fig. 96), or square in shape and arranged in rows and columns (two-dimensional array, Fig. 97), or ring-shaped and arranged concentrically (annular array, Fig. 98).

A linear switched array (sometimes called a linear sequenced or simply a linear array) is operated by applying voltage pulses to groups of elements in succession as shown in Fig. 99. In this way, the sound beam is moved across the face of the transducer electronically producing a picture similar to that obtained by scanning a single transducer manually. If the electronic sequencing or scanning is repeated fast enough (30 frames per second), a real-time image can be generated. Linear arrays are usually 1 cm wide and 10 to 15 cm long with 64 to 256 elements.

The linear phased array, while similar in construction, is quite different in operation. A phased array is smaller (1 cm wide and 1 to 3 cm long) and usually contains fewer elements (32 to 128). The ultrasonic beam generated by a phased array can be both focused and steered by properly delaying the signals going to the elements for transmission or arriving at the elements for receiving. This is illustrated in Fig. 100 and Fig. 101. Figure 100 shows how focusing can be achieved with a five-element array. The pulse exciting the center element is

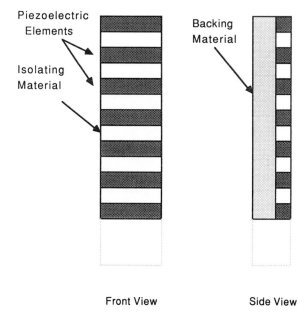

Front View                    Side View

**Figure 96**  Construction of linear array: (a) front view; (b) side view.

Piezoelectric  Elements

Isolating
Material

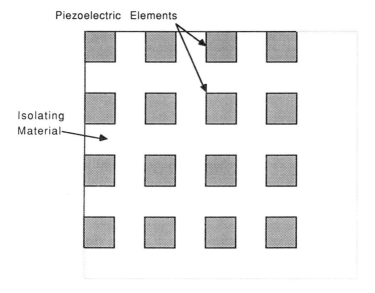

**Figure 97**  Front view of 2-D array.

Piezoelectric
Elements

Isolating
Material

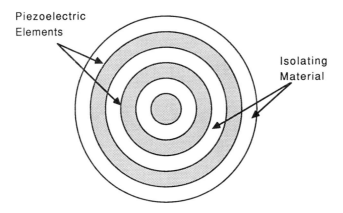

**Figure 98**  Front view of annular array.

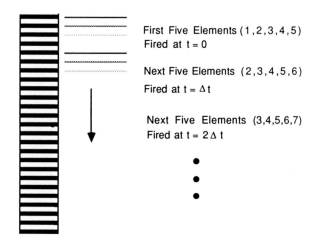

First Five Elements ( 1 ,2 ,3 ,4 ,5 )
Fired at t = 0

Next Five Elements ( 2 ,3 ,4 ,5 ,6 )
Fired at t = Δt

Next Five Elements (3,4,5,6,7)
Fired at t = 2Δt

**Figure 99**  Linear switched array produces images by successively exciting groups of piezoelectric elements.

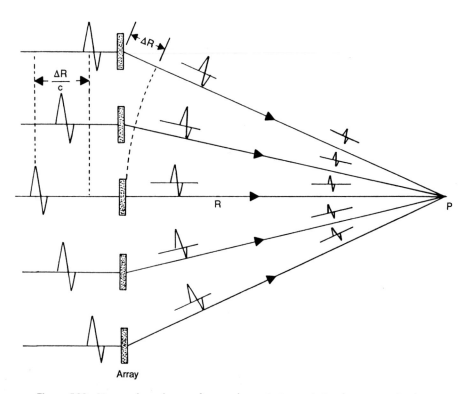

Array

**Figure 100**  Linear phased array focuses beam in transmission by appropriately delaying excitation pulses to different elements.

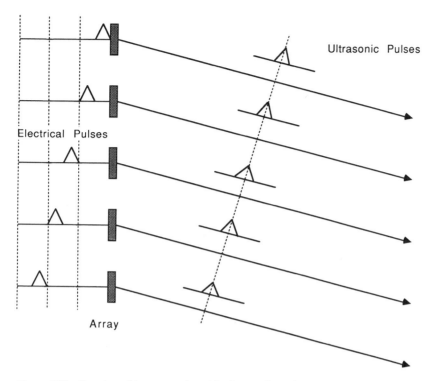

**Figure 101**   Steering of beam produced by linear phased array.

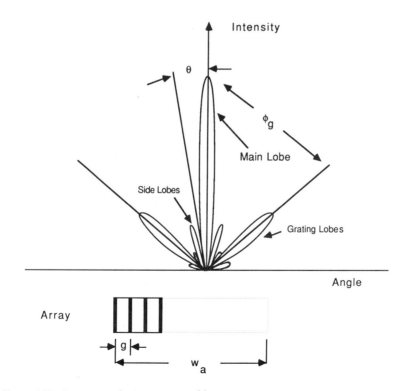

**Figure 102**   Intensity radiation pattern of linear array.

delayed by a time period $\Delta R/c$ relative to the pulses exciting the elements on the perimeter, so that all ultrasonic pulses arrive at point $P$ simultaneously. Figure 101 shows how steering in transmission can be achieved by properly sequencing the transmitted pulses. By appropriately adjusting the delays, beam steering and focusing can be produced simultaneously.

The linear arrays generally suffer from deterioration in lateral resolution because of high side-lobe energies called the grating lobes generated by regularly spaced elements (Macovski, 1979). Referring to Fig. 102, the angles at which grating lobes occur are given by

$$\phi_g = \sin^{-1}(n\lambda/g) \qquad (2\text{-}50)$$

where $n = \pm 1, \pm 2, \ldots$, and $g$ is the gap between elements. The width of the grating lobe and the main lobe is given by

$$\theta = \sin^{-1}(\lambda/w_a) \qquad (2\text{-}51)$$

where $w_a$ is the width of the array. Equation (2-51) differs from (2-48) because these elements are rectangular in shape. The magnitude of grating lobe relative to the main lobe is determined by the width of the element, $w_e$, The smaller $w_e$, the larger the magnitude of grating lobes relative to the main lobe (Christensen, 1988). There are several ways to minimize the grating lobes (Larson, 1983). The spacings between elements may be randomized or $g$ be made as small as possible. The third alternative is to make the pulses as short as possible. However, none of these approaches are perfect.

Recall that linear arrays can be focused and steered only in one plane. Focusing in the direction perpendicular to the array can only be achieved with a lens. This problem may be alleviated by using two-dimensional arrays. Two-dimensional arrays have been studied by various groups but have never been produced commercially (Wade, 1976; Smith *et al.*, 1991). A 20 × 20 2-D array has been built by Smith *et al.* to perform high-speed 3-D ultrasonic imaging.

The fourth novel probe is the annular array, as shown in Fig. 98. This type of transducer can achieve biplane focusing. By using appropriate externally controllable delay lines or dynamic focusing, focusing throughout the field of view can be attained. This type of probe cannot provide beam steering. Mechanical steering has to be used to generate 2-D images. The width of each annulus may be adjusted to maintain the beam intensity over the dynamically focused zones. The most advanced annular arrays have been fabricated of piezoelectric ceramic/polymer composites.

All forms of arrays suffer from wave diffraction artifacts beyond the focal zone. Recently promising results have been obtained for an experimental non-diffracting annular array transducer, based on the non-diffracting solution to the wave equation (Durnin, 1987), by Lu and Greenleaf (Lu and Greenleaf, 1990). Instead of driving the annuli with Gaussian shaded signals, that is, the signal amplitudes driving the annuli have a Gaussian profile, this device uses Bessel shaded driving signals.

# III.   Ultrasonic Diagnostic Methods

## A.   Pulse–Echo Systems

Current ultrasonic diagnostic imaging systems are mostly operated in the pulse–echo mode. A single probe is used both for transmitting the signal and for receiving the echoes reflected or scattered back from the tissues. To obtain the best axial resolution, the probe is excited by extremely short pulses. Depending on how the information is displayed, pulse–echo methods can be classified as A, B, C, or M modes.

## 1. A or Amplitude Mode

The oldest and simplest type of pulse–echo ultrasound instruments uses A mode display or a display of the echo amplitude as a function of depth of penetration, which is equal to $ct/2$ where $t$ is the time of flight of the pulse, on a monitor. Figure 103 shows the essential components for such a system.

The transducer is excited by the pulse generator whose output is a high-voltage pulse train with a peak-to-peak voltage of more than 100 V and a fast rise time. The pulse repetition rate is approximately 1 kHz. Since in pulse–echo systems the same transducer is used for transmission and reception, a pulse should not be transmitted until all the echoes from structures of interest are received. As a result, the maximal pulse repetition rate is limited by the depth of penetration. Otherwise range ambiguity of the exact location of the echo may result.

The ultrasound pulse may be coupled to the subject through an aqueous gel or oil (contact scan) or through a water path. Since the transmitted pulse is a high-voltage spike and the received echoes are typically very small, a low-noise preamplifier is placed ahead of the radio-frequency amplifier to serve as a buffer. The main amplifier must have very good overload capabilities, a short recovery time, and a wide dynamic range (more than 60 dB). The output from the amplifier is then rectified, filtered, and envelope-detected. This processed signal is used to drive the vertical axis of a monitor. The time base or the horizontal axis is synchronized with the pulse generator. The A mode display of an organ situated underneath the skin surface shown in Fig. 104(b) is depicted in Fig. 104(a).

A scan has been used for detecting foreign objects in the eye, for measuring the axial length of the eye, and for localization of brain midline structures (echoencephalography). The ability of the ultrasound to locate the brain midline is based on the fact that, in the normal, the midline structures of the brain lie in a

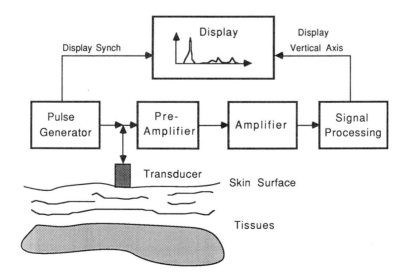

**Figure 103**  Block diagram of A-scan system.

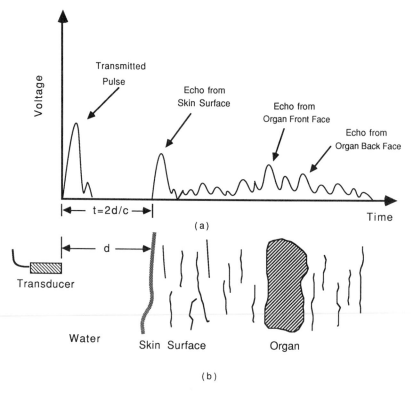

**Figure 104** (a) A-mode display of (b) organ situated beneath skin surface.

plane at the geometric center of the skull. This is at the time center of the corresponding A-scan produced by holding the probe in contact with the temporoparietal region. The midline echo can easily be seen on the A-scan because it is usually the largest intracranial echo. A-scan information has also been found useful in detecting eye tumor, liver cirrhosis, and myocardium infarction (Wells, 1977).

Historically, ultrasonic frequencies between 2 and 5 MHz are used for abdominal, brain, and cardiac examinations, and 5 to 15 MHz are used in ophthalmology, pediatrics, and peripheral blood vessel studies. These frequencies are chosen in a trade-off among resolution, power levels, and the attenuation of the intervening tissues.

**Signal Compression**  Since the specularly reflected echoes from tissue interfaces are much larger than the diffusely scattered echoes from tissue parenchyma, the dynamic range of the echo amplitude can be as high as 70 to 80 dB. The dynamic range of a device is defined as 20 log (maximal signal/minimal signal that the device can handle without significant distortion). It is not only difficult to amplify signals with such a wide dynamic range without distortion but also impossible to find a monitor that can display signals with a dynamic range greater than 40 dB. Consequently, in almost all ultrasonic imaging devices, the returned echoes detected by the probe are logarithmically compressed before amplification and further processing.

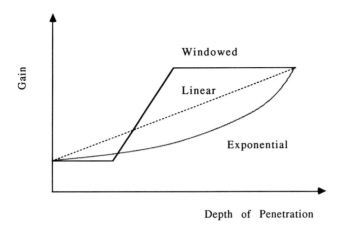

**Figure 105**   Various time-gain compensation curves that have been used in commercial scanners.

**Time-Gain Compensation**   As was discussed in Section I-G of this chapter, the intensity of the ultrasonic beam decreases exponentially as it penetrates the tissue. Also the energy of the beam is lost due to divergence of the beam. Therefore, two identical targets at different depths will produce different echoes with the echo produced by the closer target being larger than the other. This problem may be circumvented by using the time-gain compensation (TGC) in which the gain of the amplifier is increased as a function of time or depth to compensate for the loss in energy due to attenuation and beam divergence. Various forms of time-gain compensation have been used. Some of them are shown in Fig. 105. In modern scanners, TGC shape can be conveniently adjusted to optimize the image or the application. It must be realized here that there are limits on the depth of penetration and the maximal gain set by the inherent noise contained in the electronic components as well as in the acoustic components.

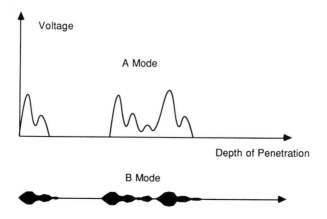

**Figure 106**   B-mode display.

## 2. B or Brightness Mode

The information carried by the A-mode display shown in Fig. 104(a) can also be represented in a format shown in Fig. 106, or the brightness mode, where the echo amplitude is used to modulate the electron beam intensity of a cathode ray tube (CRT) or the $Z$ axis of the CRT instead of the vertical or $Y$ axis. If the electron beam has only two states, on when the echo amplitude is above a certain threshold and off when it is below the threshold value, this type of B-mode display is called bistatic B mode display. If the intensity of the electron beam is made to be proportional to the echo amplitude, this type of display is called gray-scale B mode display. The display shown in Fig. 106 represents one line of information (one A-line), that is, the echo amplitude as a function of the depth of penetration in one probe direction. If more A lines can be obtained by either rotating or linearly translating the probe manually and the probes spatial location can be encoded into a voltage, a 2-D image of a structure can be formed.

The block diagram of a static B-mode scanner is shown in Fig. 107. The returned echoes from the tissues are first amplified by the preamplifier and the variable gain amplifier, which provides the time-gain compensation, and then is logarithmically compressed. The compressed signal is subsequently demodulated to detect only the envelope and filtered to remove the carrier signal. The demodulated signal and the electrical signals generated by the position sensors mounted on an articulated arm holding the transducer have to be converted into a format suitable for display in the TV format. This is accomplished by the scan converter. Both analog and digital scan converters have been used. Analog scan converters were used by and large in older machines. Modern ultasonic scanners use digital scan converters because of their flexibility and superior performance. The output of a scan converter can be further processed, for example, smoothing, filtering, thresholding, and/or gray-scale mapping to enhance the image. Signal processing

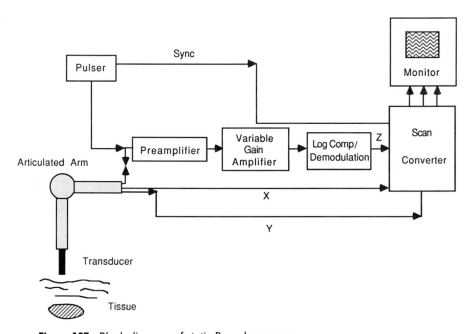

**Figure 107** Block diagram of static B-mode scanner.

that is performed after scan conversion is called postprocessing whereas what is performed before the scan conversion is called preprocessing. It is obvious that many of these functions can easily be managed by a computer, which has become an essential part of a modern ultrasonic scanner.

This type of scanner is called a static B-scanner in contrast to a real-time B-scanner, which will be discussed later, because typically several minutes are needed to construct an image. Therefore, it cannot be used to image any structures in motion. Moreover, because the reflection from flat interfaces is directional, different types of motions of the probe, (e.g., linear plus sector) may have to be combined to produce a better image. This kind of scanning procedure is called compound B scan.

**Real-Time B Scanners**    The disadvantages of the manual B-scan systems, namely, being more operator-skill dependent and time consuming, can be overcome by using real-time systems that produce images fast enough so that motions of the organ can be displayed and in which the beam is automatically steered so that they are less dependent on operator skill. There are mechanically and electronically scanned real-time systems. Currently, practical ultrasonic equipment operates on the principle that only one acoustic pulse travels in the field of interest at any instant of time. This puts a constraint on real-time scanners. The maximum frame rate, or the maximum number of images produced per second, depth of field, and number of scan lines per image are related by the equation

$$R_f \cdot D \cdot N = \frac{c}{2} \qquad (2\text{-}52)$$

where $R_f$ is maximum frame rate (sec$^{-1}$), $D$ is depth of field (m), $N$ is number of scan lines, and $c$ is velocity of sound (m/sec).

To achieve an improvement in one factor, another must be sacrificed. For instance, to obtain more scan lines in the image, either the frame rate or the depth of view must be decreased. Therefore, high-quality real-time images are difficult to obtain for large organs, such as the liver, which require a large field of view. Typically in a real-time ultrasonic image there are 128 lines. For a frame rate of 30/sec the maximal depth of view or penetration is 19.5 cm. Here it should be noted that the depth of penetration is limited by the pulse repetition frequency or rate of the pulses that excite the transducer.

The simplest real-time system is to replace the human hand with a mechanically driven device, as illustrated in Fig. 108(a). The transducer is housed in a dome made of plastic and bathed in a fluid. The transducer is rocked back and forth at a rate of more than 30 times/sec. Each sweep of the transducer produces one image. However, the oscillation of the contact scanner against the skin may be uncomfortable for the patient and at times this motion causes the images to be blurred, since the tissues may be moved by vibration. These motion artifacts can be eliminated by using an electronic scanner or an alternative design, as shown in Fig. 108(b). A rotating wheel on which four transducers are mounted is contained in a sealed, oil-filled, cylinder. This type of device can generate a 90° sector scan of four frames per revolution, which reduces considerably the speed of the rocking motion of the transducer and consequently the vibration associated with the motion.

One of the advantages of a mechanical scanner as compared to an electronically steered scanner is that it does not have the grating lobe problem, but the im-

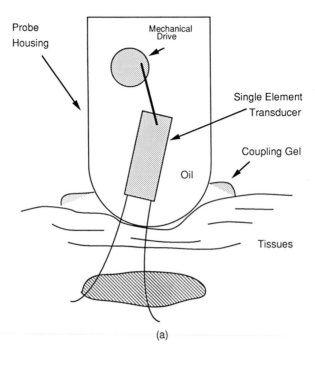

(a)

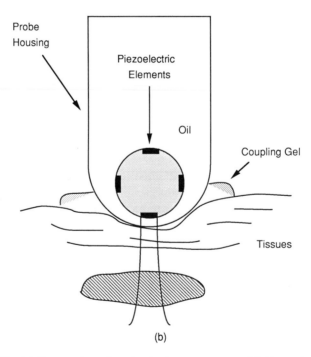

(b)

**Figure 108**   (a) Construction of mechanical sector scanner. (b) Alternative design of mechanical sector scanner.

age produced by a mechanical sector scanner is sometimes corrupted by the presence of the transducer dome and the bathing fluid. A line of mechanical sector probes can be seen in Fig. 109(a) in which the cross-shaped probe is a standalone Doppler probe for measuring blood flow. Figure 109(b) shows a real-time ultrasonic scanner being used to examine a patient.

There are three different types of electronic scanners, depending on the probe used: the linear array, the linear phased array, and the annular array. Two-dimensional arrays, based on integrated circuit technologies, are still in the experimental stage (Smith *et al.*, 1991). A linear array scanner and associated probes are shown in Fig. 110(a). An image of the face a fetus of 37-week gestational age produced by a 3.5-MHz linear array probe is shown in Fig. 110(b). The block diagram of a typical electronically steered linear array system is shown is Fig. 111. The pulser and the ramp generator are synchronized by the clock. The pulses generated by the pulser excite groups of piezoelectric elements sequentially. The switching network controls the sequencing and is activated by the pulses from the pulser. The output from the fast ramp generator sweeps the electron beam of the CRT vertically while a slower ramp at a rate of more than 30 sweeps/sec from the switching network drives the horizontal axis of the display.

The disadvantage of a linear array is its large aperture, which makes it unsuitable for applications where there is only a limited acoustic window such as cardiac examinations. Therefore, in these situations, linear phased array systems, which have a smaller aperture, are preferred. A linear phased array system is shown in Fig. 112. An image of the heart obtained by a phased array system with a 5-MHz probe showing the mitral valve is shown in Fig. 113 in which the bottom trace is the patient's EKG. The disadvantages of the phased arrays are their complexity and the presence of grating lobes in comparison to mechanical sector scanners.

The scan line density of linear arrays is uniform throughout the field of view. This is not so for sector scanners. The line density is higher near the probe but decreases farther from the probe, resulting in poor image quality in the far field. It is for this reason that interlacing between scan lines is usually necessary to smooth the image.

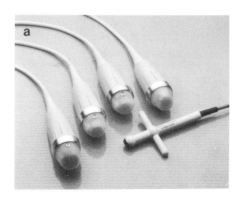

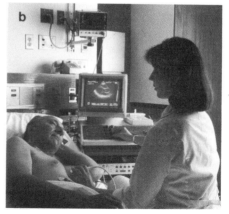

**Figure 109** (a) Mechanical sector probes. (b) Real-time mechanical sector scanner in operation. (Courtesy of Hewlett Packard, Andover, Massachusetts).

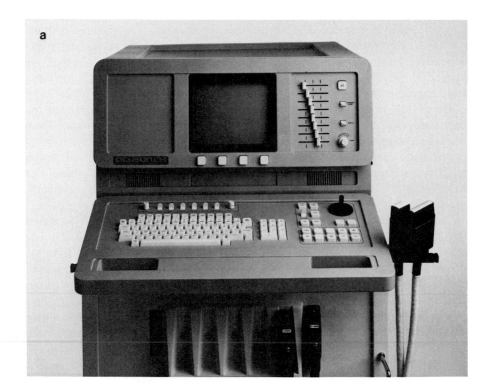

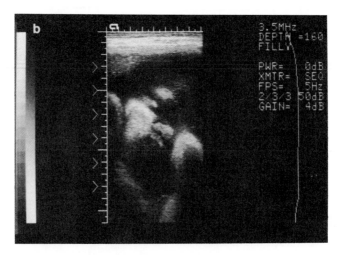

**Figure 110**   (a) Linear array scanner. (b) Image of fetus *in utero*. (Courtesy of Acuson, Mountain View, California).

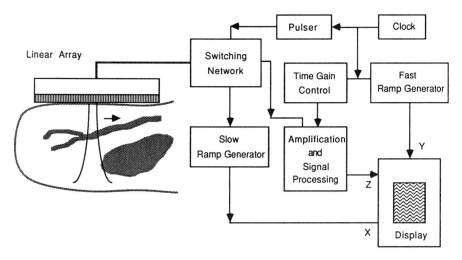

**Figure 111** Block diagram of a linear array scanner.

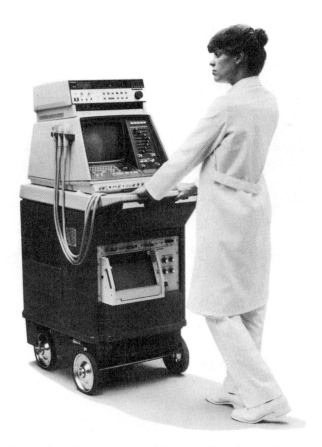

**Figure 112** Linear phased array scanner. (Courtesy of Hewlett Packard, Andover, Massachusetts).

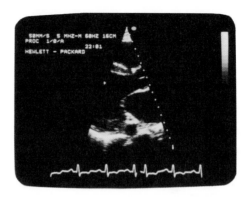

**Figure 113** Image of mitral valve obtained by phased array. (Courtesy of Hewlett Packard, Andover, Massachusetts).

Commercial real-time scanners are available with frequencies ranging from 2 to 10 MHz depending on the applications. High-frequency scanners can only be used for imaging smaller organs like thyroid, testicles, etc., and are also called small-part scanners.

B-mode imaging has been used in a variety of clinical problems. The most well known are fetal monitoring in obstetrics, diagnosing masses in abdominal organs such as liver and kidney, and assessing cardiac conditions, to name just a few.

## 3. M or Motion Mode

M mode uses a standard A-mode instrument with a modified display. The M-mode display resembles the B mode with the exception that the horizontal or $X$ axis is driven by a very slow ramp at a rate of less than 1 sweep per second. This is shown in Fig. 114 where the $Y$ axis indicates the distance of the echo from the transducer and the $X$ axis indicates the time. Figure 114 illustrates the M-mode

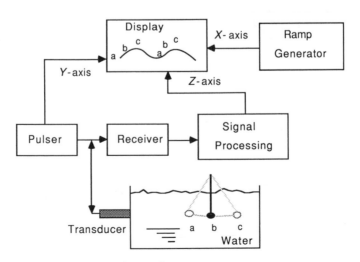

**Figure 114** Block diagram of M-mode system.

display of a moving pendulum. The trace displayed delineates the position of the echo as a function of time. M-mode display is useful for mapping the movement of moving objects such as mitral valve in the heart. An M-mode display of heart structures is shown in Fig. 115(b). Figure 115(a) shows the B-mode image of the heart obtained by a mechanical sector at 5 MHz. The middle dotted line indicates the direction of the tranducer along which the M-mode display was obtained.

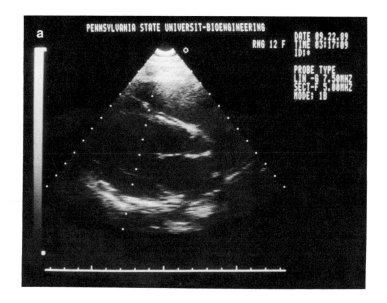

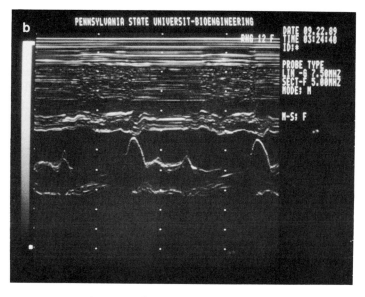

**Figure 115**   (a) B-mode image of mitral valve. (b) M-mode display of mitral valve motion along direction indicated by middle dotted line in (a).

## 4.   C or Constant-Depth Mode

C mode instrumentation is very similar to that of B mode. However, unlike B mode, where one dimension of the display is inferred from the time-of-flight information of the ultrasonic pulse, the scanning plane is fixed at a constant depth from the probe in C-mode imaging. The selection of a certain plane for C-mode imaging is achieved by time gating of the returned echo waveform. The time-gating circuit functions like a switch that is turned off with no triggering pulse but can be turned on by a pulse for a time duration controlled by the pulse length. To obtain C-mode imaging, a time gate triggered by the pulser delayed by a time period equivalent to the depth may be inserted between the variable-gain amplifier and the demodulator. In this case, the $X$–$Y$ plane would represent a plane parallel to the scanning plane of the transducer and perpendicular to the ultrasonic beam, whereas in B mode the $X$–$Y$ plane being imaged is a plane containing the ultrasonic beam. C-mode imaging has not been found useful in clinical diagnosis but occupies an important position in acoustic microscopy (Wade, 1976). Scanning acoustic microscopes (SAM) with frequencies ranging from 100 MHz up to 3 GHz have been developed, and at 3 GHz their spatial resolution is as good as an optical microscope with the advantage that ultrasound can penetrate light opaque material.

## B.   Transmission Methods

In contrast to pulse–echo systems, which use one transducer for transmitting and receiving ultrasound, a second transducer is used for receiving the transmitted pulses in ultrasonic transmission imaging methods. In this approach, ultrasonic attenuation along the propagation path is measured and the image has little dependence on the backscatter or reflecting properties of the tissue, which are the major determinants of pulse–echo images. Methods that measure transmitted ultrasound include transmission-scanning acoustic microscopy, laser-scanning acoustic microscopy (Kessler, 1985), and ultrasound computed tomography (Greenleaf, 1983).

### 1.   Scanning Laser Acoustic Microscopy

In the scanning laser acoustic microscope (SLAM), a specimen is viewed by placing it on a crystal stage where it is insonified with plane ultrasonic waves and illuminated with laser light that senses the minute displacements on the specimen surface or on the surface of a cover slip placed on top of the specimen, which occur as a sound wave propagates through them. The disturbances are dynamic in that the pressure wave is periodic and the surface displacements accurately follow the wave amplitude and phase. By electronically magnifying the area of laser scan to the size of the CRT monitor, and by brightness-modulating the display, the acoustic micrograph is made visible. If the sample is made of a solid substance that can be optically polished, the sample is viewed directly with laser. However if the sample is not polished (for example, biological tissues), a plastic mirror (cover slip) is placed in contact with the sample to relay the acoustic information to the laser beam.

### 2.   Computed Tomography

The basic principles for ultrasonic computed tomography are similar to those of X-ray tomography. The gantry consists of either a single transducer as a transmitter and a single transducer as detector or arrays as transmitters and detectors. Two types of ultrasonic computed tomograms can be generated: one that depicts ultrasound velocity and the other that depicts attenuation. The quality of ultrasound CT images has been poor in comparison to B-mode images, probably for two reasons, beam refraction at tissue interfaces and beam divergence, which are difficult to correct. Moreover since ultrasound cannot penetrate the whole body nor organs like lung and bone, without excessive patient radiation the potential applications of ultrasound CT are likely to be limited to small organs like breast and testicles.

## C. Doppler Methods

The Doppler effect discussed in Section I-I can be used to detect tissue or organ movement, or blood flow in blood vessels. Ultrasonic Doppler instruments are currently used for monitoring the fetal heartbeat, detecting multipregnancies, measuring cardiac output, and diagnosing stenosis in blood vessels (Hatle and Angelsen, 1985).

There are basically two types of ultrasonic Doppler systems, namely the continuous-wave (CW) Doppler system, and the pulsed-wave (PW) Doppler system. The basic physical principle for the CW Doppler system is very simple. Assuming that a Doppler probe consists of two elements, one the transmitting crystal and the other the receiving crystal, and the ultrasonic beams are making an angle $\theta$ with the direction of blood flow, as shown in Fig. 116, the Doppler frequency shift can be calculated using Eq. (2-45).

$$f_d = \frac{2v \cos \theta}{c} f \tag{2-53}$$

where $f_d$ is Doppler-shift frequency, $c$ is the ultrasound velocity in blood, $f$ is the ultrasound frequency, and $v$ is the blood flow velocity.

This equation shows that Doppler shift is proportional to the frequency of incident ultrasound as well as the velocity of the flow. Assuming a flow speed of $v = 10$ cm/sec and an ultrasonic frequency of 5 MHz, an ultrasonic speed of $1.5 \times 10^5$ cm/sec, $f_d = 470$ Hz, which is in the audio range.

A block diagram of a continuous-wave Doppler system is illustrated in Fig. 117. The received signal is amplified and mixed with the reference frequency. The output of the demodulator contains the Doppler frequency. After going through a low-pass filter and an audio amplifier, the signal can be heard from a speaker. Alternatively, the mean frequency of the audio signal is estimated from the zero-crossing frequency of the waveform, which is related to the mean frequency (Atkinson and Woodcock, 1982) and can be measured by a zero-crossing

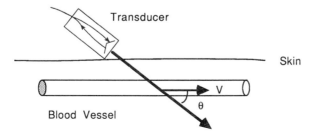

**Figure 116**  CW Doppler method for measuring blood flow velocity.

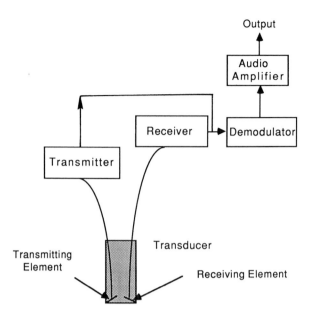

**Figure 117** Block diagram of CW ultrasonic flowmeter.

counter. Frequently, a spectral analyzer is used to obtain the frequency spectrum of the Doppler signal. The bandwidth of the audio frequency (AF) amplifier is usually limited from 200 Hz to 10 kHz. This low-frequency cutoff is required to reduce noise and large-amplitude low-frequency signals caused by slow-moving structures like blood vessel walls. Figure 118(a) and (b) show respectively the mean flow velocity in a normal radial artery and a gray-scale spectrasonogram where the instantaneous Doppler spectrum is plotted as a function of time. In Fig. 118(b) the vertical axis indicates the Doppler frequency or velocity whereas the horizontal axis indicates time, and the magnitude of the spectrum is represented by gray scale.

Simple continuous-wave Doppler systems are unable to distinguish where the echoes come from. This may become a serious problem when there are overlapping vessels. Under these situations, pulsed Doppler systems may be used. These systems produce short coherent sinusoidal bursts. The returned echoes are range gated and demodulated to obtain the Doppler shift. The range gate indicates the location from where the Doppler shift is generated.

A block diagram of such a system is shown in Fig. 119. The pulse generator puts out a series of pulses with a pulse repetition frequency (PRF) typically in the order of a few kilohertz, which modulates the oscillator. The returned echoes from the tissues, blood, and blood vessels are amplified and gated. The time delay of the gate is adjustable. The gated RF waveform is then demodulated. A sample-and-hold device is used to sample the output from the demodulator. Following amplification and low-pass filtering, the Doppler signal can be processed as previously described. There is an upper limit to the Doppler frequency shift that can be detected by this approach. It depends on the sampling rate, which is the pulse repetition frequency. To avoid the range ambiguity, which means that the echoes from areas of interest have to be received before the next pulse is transmitted, the PRF is limited by the depth of penetration. It has been shown that

$$v_m \cdot z_m \lesssim c^2/(8f) \tag{2-54}$$

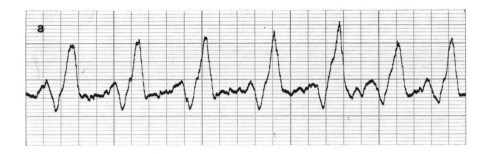

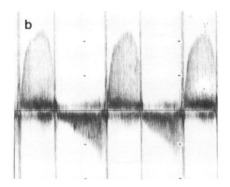

**Figure 118** (a) Mean blood velocity in radial artery measured by CW ultrasonic flowmeter. One major division on the ordinate represents velocity of 10 cm/sec. Chart recorder speed was 2.5 cm/sec (one major division on the abscissa represents 0.2 sec). (b) Spectrasonogram in which spectrum of Doppler signal is displayed every 5 to 10 msec in gray scale.

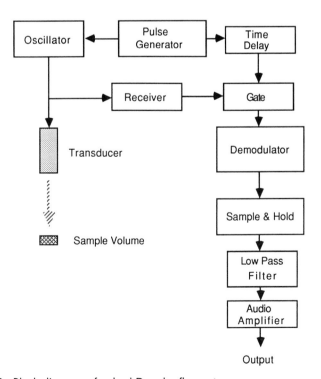

**Figure 119** Block diagram of pulsed Doppler flowmeter.

where $v_m$ and $z_m$ are respectively the maximal velocity that can be detected by a pulsed Doppler device and the maximal depth of penetration (Atkinson and Woodcock, 1982). Since the right-hand side of the equation is a constant, this equation says an increase in $v_m$ is accompanied by a decrease in $z_m$ and vice versa.

## D.  Duplex Imaging

Duplex imaging, which combines B-mode real-time imaging with ultrasonic Doppler blood flowmetry, provides unique capabilities unattainable with other imaging modalities. The B-mode can be either sector or linear and the Doppler can be either a continuous wave or a pulsed wave or both. A cursor line is typically superimposed on the B-mode image to indicate the direction of the Doppler beam. The Doppler probe is integrated with the imaging probe in some systems, while in others it is attached to the imaging probe externally. In the Hewlett-Packard phased array 77020A scanner, the 64-element phased array probe is used to extract PW Doppler information and a separate probe is used to obtain CW Doppler signal. Acquisition of the image and Doppler signal is synchronized at 120 or 180 MHz generated by a timer card, which is much higher than the 60-MHz rate of the imaging system master clock. Doppler echoes are received by the scanner at 2.5, 3.5, or 5 MHz depending on the probe used. In the Duplex mode, the scanner generates imaging and Doppler lines alternatively, thus producing an image and Doppler outputs simultaneously. The disadvantage of doing so is that the Doppler is sampled at a fairly low rate, 2 kHz, which would limit the maximum flow velocity resolution. To alleviate this problem, the image may be frozen and all lines then used for acquiring Doppler. An FFT algorithm is used to generate the real-time Doppler spectrum. Each line in the spectrum is generated in 10 msec or so.

Duplex scanning has been found to be of great clinical significance in cardiology, vascular surgery, and pediatrics. In adult cardiology, it is used for diagnosis of vascular malfunctions, assessment of cardiac functions, and detection of congenital heart diseases. In vascular surgery, duplex scanners are used to assess peripheral blood vessels. Stenosis and atherosclerosis of vessel walls can be examined with scanners operated at frequencies higher than 5 MHz.

## E.  Tissue Characterization

A significant feature of ultrasonic imaging is that quantitative characterization of the state of biological tissues by ultrasound may be possible because of the complicated manner that ultrasound interacts with tissue, as was discussed in Section I of this chapter. This was recognized as early as the 1970s when gray-scale ultrasound, in which not only the echoes reflected from tissue boundaries but also the diffusely scattered echoes from tissue parenchyma were displayed, was gaining popularity. To a certain extent, gray-scale images acquired by clinical pulse–echo scanners can provide some information for tissue characterization based on the textural pattern and the gray level (echogenicity) of the tissue, although only the amplitude of echoes scattered or reflected back from the tissues is used to form an image. For instance, different tissues can be differentiated from their textural appearance or speckle pattern. A fluid-containing cyst is usually less echogenic than a solid tumor.

Since the early 1970s, there has been an ongoing interest in developing quantitative methods for extracting additional data from the returned echoes for tissue characterization. Its ultimate goal is that these additional data may yield useful clues for tissue differentiation when the difference is not apparent in a conventional image. A number of ultrasound propagation parameters have been used for tissue characterization *in vivo*. These include velocity, absorption or attenuation, backscatter, and the nonlinear parameter $B/A$ (Greenleaf, 1986). In the following, these methods and preliminary experimental results will be discussed.

## 1.  Velocity

*In vitro* experimental results on ultrasound velocity obtained with the time-of-flight approach have indicated that ultrasound velocity in tissues can be varied by pathology (Shung, 1987). However, time-of-flight velocity measurements are difficult if not impossible to perform on humans *in vivo*. Ultrasound computed tomography is a possibility but it suffers from a number of drawbacks, including refraction and divergence of the beam. Therefore, a number of novel pulse–echo approaches have been developed.

Among the most promising is the image misregistration method (Chen *et al.*, 1987). This method, utilizing a commercial B-mode scanner to measure ultrasound velocity in liver, is diagrammatically illustrated in Fig. 120. A suitable target such as a blood vessel is first located by the scanner with the probe placed at position 1. The image of the target actually is displaced from the true location because the assumed velocity $c_0$ in the scanner differs from the true velocity $c$, which is the value to be measured. In commercial scanners, a $c_0$ of 1540 m/sec is usually assumed. The probe is then placed at position 2 and an image of the target is obtained. The true velocity $c$ is then related to $c_0$ by the following equation:

$$c/c_0 = [D/(D - d)] \tag{2-55}$$

Since the target selected generally has to be specular in nature for it to be isolated from the parenchymal echoes and the axial resolution of the probe is better than

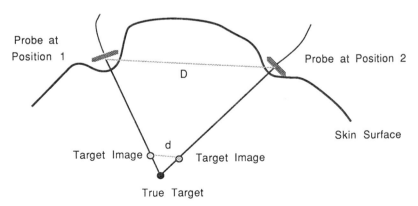

**Figure 120**  Schematic diagram illustrating method for measuring ultrasonic velocity *in vivo* based on image misregistration.

the lateral resolution, the target may not be recognizable when the probe is moved to a different position. As a result, a cross-correlation technique has to be used to match the images.

This method has been used to study a large patient population. The results show that ultrasound velocity in fatty liver (1547 m/s) is significantly lower than that of normal liver (1578 m/s). The data for normal liver are in good agreement with those obtained *in vitro*.

Other methods have also been found to be useful for estimating velocity *in vivo* (Shung, 1987).

## 2.  Attenuation or Absorption

The effect of disease on ultrasound attenuation in biological tissues has been examined extensively (Greenleaf, 1986). It has long been realized that the most likely organs for which quantitating attenuation coefficient may contribute would be large organs like liver or spleen where reasonable results can be obtained by averaging the data over a large volume of tissue.

Traditional transmission approaches for attenuation measurements, which use two probes, one as a transmitter and one as a receiver, are difficult to implement *in vivo*. Novel *in vivo* methods can be classified into two categories. A concept common to both approaches is treating tissue as a uniform distribution of scatterers. If the difference in transducer beam characteristics at two different depths in a tissue can be adequately compensated, the difference in echo amplitude from same volume of tissues located at different depths is related to only the attenuation coefficient of the tissue and the distance between the two depths.

**Loss of Amplitude Method**    This method uses the echo amplitude drop as a function of penetration depth in A mode to estimate the attenuation coefficient.

**Frequency Shift Method**    The frequency shift method (Kuc, 1984) can be accomplished either in the frequency domain (measuring spectral difference) or in the time domain (measuring zero crossing). The spectral difference approach is illustrated in Fig. 121. A-line echoes from two regions of tissues at different depths are selected by time gating. Corresponding spectra following Fourier transform are obtained. To minimize the effect of the gating pulse on the echo spectrum, Hanning or Blackman window can be utilized. The two spectra are then divided to yield the slope of the attenuation curve. In logarithmic scale, the difference in spectra yields the attenuation coefficient as a function of frequency. This process has to be repeated for hundreds of A lines before any meaningful data can result because biological tissues are far from being homogeneous.

This shift in frequency can also be estimated directly from the center frequencies of the ultrasound pulse at different depths by determining the zero crossing of the waveforms at these depths. Both of these techniques have been used for *in vivo* measurements of attenuation coefficient in liver. The normal mean value for liver has been found to be from 0.45 to 0.5 dB/cm-MHz whereas the value for fatty liver is from 0.60 to 0.80 dB/cm-MHz, which is significantly higher.

## 3.  Scattering

Since an ultrasonic image is formed from both specularly reflected echoes by tissue boundaries and diffusely scattered echoes by tissue parenchyma, it seems that a measurement of intrinsic scattering properties of a tissue is more likely to yield

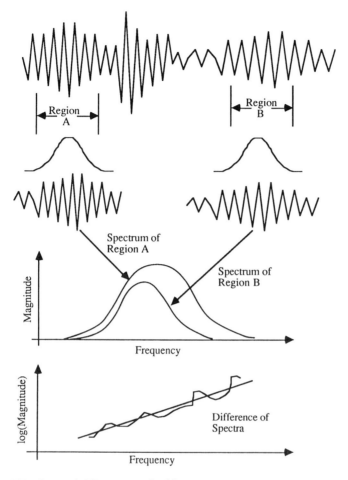

**Figure 121** Spectral difference method for estimating attenuation *in vivo.*

diagnostically more useful information than other parameters. By merely quantitating the gray levels or echogenicity of a tissue in video images acquired by a scanner following appropriate standardization of scanning procedures, a number of studies have shown that it is possible to differentiate abnormal tissue from normal tissue in a variety of organs. More quantitative data can be retrieved by acquiring and analyzing the RF signals. These data include backscatter coefficient defined as the power scattered in the backward direction by a unit volume of scatterers per unit incident intensity, frequency dependence of backscatter, and integrated backscatter, an average of the backscatter signal over the frequency band, which is less sensitive to phase cancellation artifacts and represents a more localized property.

A system based on a modified real-time scanner capable of measuring integrated backscatter in real time has been developed and is being used on patients for the diagnosis of a variety of heart problems (Milunski, 1989). A typical result obtained from a patient is shown in Fig. 122, where the top trace indicates the cyclic variation in integrated backscatter in decibels whereas the bottom image shows the M-mode display with the site of analysis in the septum being indicated

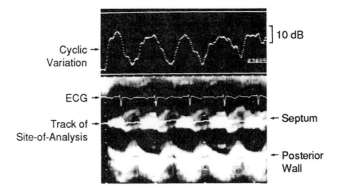

**Figure 122** Top panel depicts integrated backscatter from septal myocardium whereas lower panel shows M-mode display. (Courtesy of James Miller, Washington University, St. Louis, Missouri).

by the bright line. *in vivo* tissue characterization with scattering has also been pursued on other tissues, most notably the eye and liver (Feleppa and Lizzi, 1986). Tissue characterization occupies a particularly important place in ophthalmology because biopsies are not possible in the eye.

In general it has been accepted that the most effective way of characterizing tissue is by using a multiparameter approach, not simply relying on one parameter alone for tissue differentiation.

These results are certainly encouraging. However, there are also technical problems that have to be overcome before quantitative tissue characterization can become a clinically accepted tool. Among them are (1) how the effects of intervening tissues can be corrected in a rational manner, (2) how to better understand the fundamental physical mechanisms involved in producing these changes, and (3) how to correct transducer diffraction effects.

# IV. New Developments

## A. Color Doppler Flow Imaging

Various types of ultrasonic Doppler imaging systems produce images of blood vessel lumen (Atkinson and Wookcock, 1982). The CW Doppler imaging device operates much like a C-scan unit with the exception that the Doppler signal from the blood is detected. The amplitude of the Doppler signal is used to turn on the $Z$ axis of a CRT, while the positional information of the tranducer is loaded to the $X$, $Y$ axes of the CRT via displacement transducers. The pulsed Doppler imaging system then behaves like a B-scan unit. The position of the gate provides information about the Doppler signal in the direction of the ultrasonic beam. The position of the gate can be adjusted manually or multigates can be used to shorten the time needed to acquire an image. Usually, these images are displayed in a biostatic mode to simply indicate whether there is flow and then the flow is assessed by a separate Doppler unit. However, one major drawback of these devices is that the images are not displayed in real time.

Usually the time needed to produce an image is quite long, in the order of a few minutes. In a color Doppler flow system, the blood flow data are displayed in real time along with a gray scale B-mode image. Each A line is divided into many segments, typically more than 500. The Doppler shift is extracted from all these segments by successively transmitting pulses (eight or more) in the same direction, and the number of pulses transmitted determines the accuracy of Doppler estimation in an algorithm that uses autocorrelation calculation (Kasai et al., 1985) to estimate Doppler shift. The autocorrelation function of a time function, $f(t)$, is related to its power spectrum by the following equation.

$$R(\tau) = \int_{\infty}^{-\infty} P(\omega)e^{j\omega\tau} \, d\omega \qquad (2\text{-}56)$$

where $R(\tau)$ and $P(\omega)$ are respectively the autocorrelation function and power spectrum of the function $f(t)$. It can be easily shown that the mean frequency $<\omega>$ and variance of the power spectrum $\sigma^2$ are related to the autocorrelation function by:

$$j<\omega> = \dot{R}(0)/R(0) \qquad (2\text{-}57)$$

$$\sigma^2 = [\dot{R}(0)/R(0)]^2 - \ddot{R}(0)/R(0) \qquad (2\text{-}58)$$

Therefore, the mean velocity and variance of blood flow can be calculated from these simple mathematical equations once the autocorrelation function is known. In practice, the autocorrelation function is estimated from correlating successive A lines. It has also been reported that in at least one system, an FFT algorithm is used to obtain Doppler shift.

In a way, the color Doppler flow device is much like a multigate pulse Doppler flowmeter, but the sweep of the beam is electronically controlled. The position of the gate provides the spatial information of the flow and the Doppler shift of the signal originating from this location yields the flow information.

The block diagram of a system utilizing the autocorrelation approach is shown in Fig. 123 (Kasai et al., 1985). A probe is used for simultaneous acquisition of B-mode and Doppler data. A third data path is needed for acquiring conventional Doppler data. The B-mode image is displayed in gray scale whereas the Doppler shift is displayed in color. Conventionally, a color such as red is assigned to flow toward the transducer and a color like blue is assigned to flow away from the transducer. Magnitude of the velocity is indicated by different shades of the color (eight or more). In other words, the higher the velocity, the lighter the shade. In most color Doppler mapping systems, a third color, yellow or green, is used to indicate variance.

A color Doppler image of an internal carotid artery with mild calcified plaque obtained by a linear array is shown in Fig. 124 (see color plate 1). The shadow in the bottom part of the image indicates attenuation of the beam by calcification. The change in color indicates turbulence resulting from the presence of the plaque. A color Doppler image of the heart obtained by a phased array is shown in Fig. 125 (see color plate 2). To eliminate aliasing of Doppler estimation due to the limit imposed by the pulse repetition rate and to maintain the frame rate, a smaller area of the image may be specified as indicated by the parallelogram in Fig. 124.

Many potential clinical applications have been found for this device (Kisslo et al., 1988). Its efficacy is still being evaluated. Preliminary results seem to indicate that it is useful for diagnosing tiny shunts in heart wall and valvular regurgitation and stenosis. It reduces considerably the examination time in many diseases

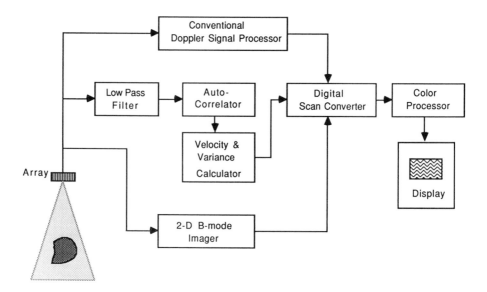

**Figure 123**   Block diagram of color Doppler flow mapping system.

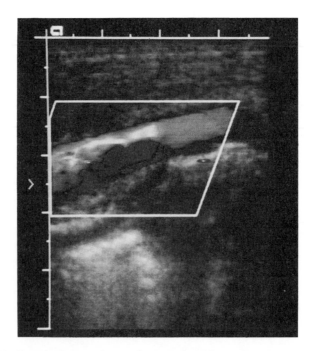

**Figure 124**   Doppler image of carotid artery where flow reversal is seen due to presence of atherosclerotic lesion in center of image. For color Doppler image see Plate 1. (Courtesy of Acuson, Mountain View, California).

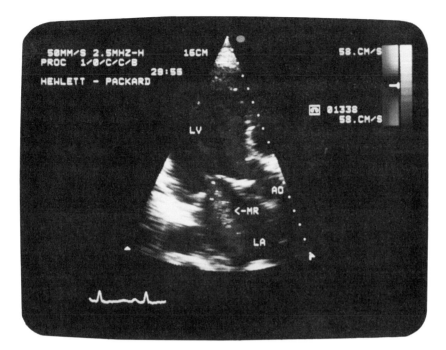

**Figure 125** Doppler flow image of left heart showing mitral valve regurgitation. For color Doppler image see Plate 2. (Courtesy of Hewlett Packard, Andover, Massachusetts).

associated with flow disturbance. Problematic regions can be quickly identified first from the flow mapping. More quantitative conventional Doppler measurements are subsequently made on these areas.

## B. Parallel Processing

Only one acoustic pulse is propagated in the tissue at any one time in conventional ultrasonic image processing. The scanning time can be considerably shortened if the acoustic pulse can be sent in several directions simultaneously, or several scan lines of information can be collected at the same time. As a result, image quality can be improved from an increase in scan-line density, or the depth of penetration can be increased while maintaining the same image quality. Moreover, patient exposure can be reduced because of the shortened scan time.

A number of approaches have been used to improve ultrasonic data acquisition rates using parallel processing schemes. One of the approaches has been to acquire data in four directions for each transmitted pulse with a wider than normal ultrasonic beam, using a phased array system (Shattuck *et al.*, 1984). This arrangement is shown in Fig. 126, where the dashed lines indicate the transmitted beam and the solid lines indicate the received beams. To form the received beams, tapped delays are added to the outputs of the main delay system for each receive channel. Display of the received signals is accomplished by a single CRT using a form shown in Fig. 127. The position of the electronic beam is indicated by the zigzag line. The beam is turned on when it is in the vicinity of one of the four scan directions. Initial testing of the system with the AIUM test object showed that the resolution is not compromised by the increased frame rate, although a small reduction in the signal-to-noise ratio is expected from using a

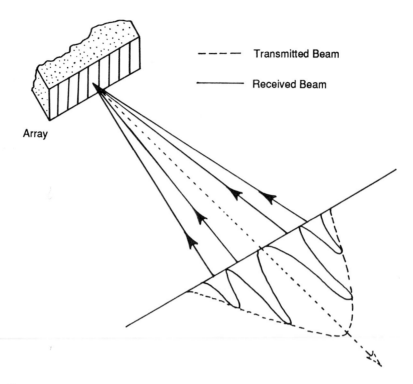

Array

– – – – Transmitted Beam

———— Received Beam

**Figure 126**   Scheme for parallel ultrasonic signal acquisition. (Courtesy of David Shattuck, University of Houston, Houston, Texas).

wider beam. Another benefit of the approach is that the distracting spacing between scan lines is removed without artificial interlacing.

A prototype high-speed volumetric imaging system has been developed based on the parallel processing concept described above (von Ramm *et al.*, 1991). Eight-line parallel processing in the receive mode is achieved with a 2-D array of 289 elements (17x17 square array). The system currently can process 4992 scan lines at a rate of approximately eight frames per second.

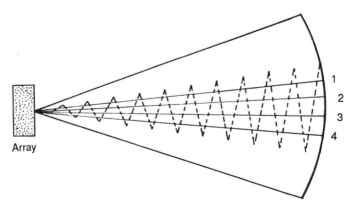

Array

1
2
3
4

**Figure 127**   Scheme used for displaying four A lines on monitor. (Courtesy of David Shattuck, University of Houston, Houston, Texas).

Parallel processing certainly increases the complexity of an ultrasonic imaging system. However, the benefits seem to far outweigh the pitfalls.

## C.  Ultrasound Contrast Media

Contrast media have been used extensively in X-ray to enhance certain structures in an image. Various types of contrast media have also been developed for ultrasonic imaging (Meltzer and Roelandt, 1982). A majority of them utilize micro-airbubbles generated as a result of chemical action or mechanical agitation. The present problems involved are producing bubbles of suitable size and bubble lifetime. In cardiology, it is preferable that the contrast medium be injected intravenously to visualize the left heart. It is therefore necessary for a significant number of bubbles to be able to travel through the pulmonary circulation. Only bubbles of less than 10 $\mu$m are capable of passing through the pulmonary capillaries.

The requirement for the bubbles to maintain their size stems from the fact that their size has to be known or maintained when being imaged or monitored in order to quantitate cardiac perfusion from two-dimensional echocardiography. Agents that have been investigated for this purpose include gelatin-encapsulated bubbles, agitated and sonicated 0.9% saline, Renografin-76, hydrogen peroxide, albumin-encapsulated bubbles, and galactose microparticles (Shapiro *et al.*, 1989). Among them the galactose microparticles and the albumin-encapsulated micro-airbubbles seem the most promising. These agents produce bubbles 3 $\mu$m in diameter when suspended in a carrier medium, and particles remain stable for at least a few minutes. Since these agents, shown to be nontoxic, contain air, they are highly echogenic even at low concentrations and are now under clinical evaluation for quantitating cardiac perfusion and better visualization of cardiac chambers (Feinstein *et al.*, 1990). Solid contrast particles have also been investigated for enhancing echo contrast of organs such as the liver (Parker *et al.*, 1987).

## D.  Intracavity Imaging

Conventional ultrasonic imaging of internal organs over the body surface is limited by the attenuation of the ultrasound energy imposed by the intervening tissues between the probe and the organ of interest. Although it is known that higher frequencies yield better resolution, a compromise has to be reached between concern with patient exposure level and resolution. Intracavity scanning offers an alternative. Better images of the heart and part of the heart (e.g., base of the heart, which is usually difficult to assess by conventional scanning) can be obtained by transesophageal echocardiography (deBruijn and Clements, 1987). The novel part of the system lies in the design of the phased array probe, which has to be small for comfort insertion, easily manipulated to achieve the best viewing angle, and adequately insulated to prevent electric shock to the patient. Additional advantages of transesophageal scanning are that the probe can be positioned in place for a long period of time for continuous monitoring of the cardiac functions during a surgical procedure and that it can be used to image obese patients and patients with lung problems or abnormal chest configurations.

Novel probes in the form of mechanical sector, linear array, or phased array are also available for transvaginal and transrectal scanning. Because of the proximity of the probe to the organ of interest, higher-frequency probes that allow better resolution can be used. Transvaginal and transrectal probes have been found useful for monitoring the fetus in early stages of pregnancy and for diagnosing prostate tumor, respectively.

Intravascular imaging, in which a transducer at a frequency typically higher than 10 MHz is mounted on the tip of a catheter for visualizing blood vessel lumen and vessel wall, is also attracting growing interest. It has been postulated that surface roughness of atherosclerotic plaque is one of the causes that lead to embolization and at present there are no reliable techniques that can satisfactorily depict lumen surface characteristics; therefore, intravascular ultrasound appears to be a viable approach (Yock *et al.*, 1988). Figure 128 (a) and (b) show respec-

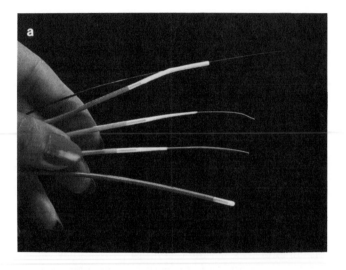

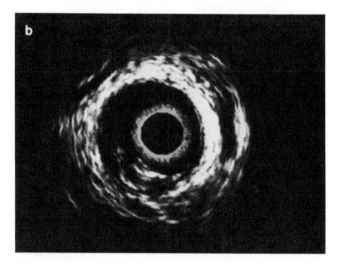

**Figure 128** (a) Various catheterization devices. Intravascular 20-MHz mechanically rotated probe mounted on 6F catheter has plastic cap at tip. (b) Image of iliac artery obtained by intravascular probe. Inner and outer echogenic rings are probe surface and blood-vessel wall interface. A plaque is seen in middle part of image at 7 o'clock direction. (Courtesy of Diasonics, Milpitas, California).

tively the catheter-mounted probe and a corresponding image of a complex atherosclerotic plaque in the posterior wall of an iliac artery obtained with the probe. In Fig. 128 (b) the dark region is blood which is less echogenic and the bright ring is the arterial wall.

The discussions given above are intended to highlight some of the most important developments in ultrasonic imaging which is still progressing at a rapid pace. There are other important and interesting advances. Among them 3-D ultrasonic image reconstruction (McCann et al, 1988), high-frequency imaging system development, applications and tissue characterization (Lockwood, *et al.*, 1991), imaging of tissue properties and motion (Insana and Hall 1990; Ophir *et al.*, 1991), and time-domain blood flowmetry (Bonnefous and Pesque, 1986; Embree and O'Brien, 1990), and blood flow estimation from frame-to-frame correlation of B-mode images (Trahey *et al.*, 1988). Three-dimensional reconstruction in ultrasonic imaging is still in its infancy. Its application in cardiology, obstetrics, and other clinical disciplines, and various methods for better visualization of 3-D ultrasonic images, is being explored. Imaging systems and probes of frequencies ranging from 20–80 MHz are being developed and studied for imaging superficial structures of the eye and skin and for intravascular imaging of blood vessel wall and blood. Time-domain flowmetry in contrast to Doppler flowmetry does not use Doppler frequency shift to extract the flow information and instead uses a correlation approach to extract the blood flow velocity component along the direction of the transducer from RF echoes scattered back by the red cells. This method has the advantages that it does not have the upper frequency limit of the Doppler devices and that it is capable of yielding 2-D flow data. It is no wonder that time-domain flowmetry is being implemented in at least one of state-of-the-art ultrasonic scanners. Alternatively, angle-independent blood flow estimation can be achieved from frame-to-frame correlation of B-mode images. At present the method is limited by poor signal-to-noise ratio and frame rate of the imagers.

# V. Image Characteristics

## A. Ultrasonic Texture or Speckle

Ultrasonic textural pattern or speckle in B-mode images, which much resembles the laser speckle, has been used routinely for clinical diagnosis. However, the origin of these patterns is still poorly understood, although it is believed to be the result of a wave interference phenomenon of the echoes arriving at the transducer from different parts of the tissue and to be related to both tissue properties and the system characteristics of the imager (Flax *et al.*, 1981). On the other end, various schemes have been implemented to smooth out the pattern in attempting to improve the ability of the scanner to resolve small objects. Present knowledge appears to indicate that tissue texture is determined both by the microstructure of the tissue and by the characteristics of the imaging device such as transducer, frequency, and bandwidth. An analysis of the texture characteristics both from computer simulation and experiments clearly demonstrates that the echoes seem to resemble laser speckle resulting from a random interference phenomenon, but embedded in the random signal there is a coherent part (Wagner *et al.*, 1983). Moreover, there is a fundamental difference between laser speckle and ultrasonic speckle in that ultrasonic signals used in imaging systems are only partially coherent, unlike coherent laser speckle, so that the speckle patterns do contain information related to tissue structure (Dainty, 1984).

## B.  Speckle Reduction

Since the presence of speckle may obscure smaller structures, thus degrading the spatial resolution of an ultrasonic imaging device, various schemes have been used to reduce the speckle appearance of an image. Most notable are frequency compounding techniques and spatial compounding techniques, both of which are based on the principle of incoherent averaging of images with different speckle patterns (Magnin *et al.*, 1982; Trahey *et al.*, 1986). In frequency compounding, images collected at different frequencies are averaged whereas in spatial compounding, images collected at slightly different spatial locations are averaged. Understandably, these speckle-reduction algorithms would lengthen the image acquisition time and also slightly affect the spatial resolution. Consequently, they have not been widely implemented on commercial scanners.

## C.  Compensation of Phase Aberration

Although in a majority of commercial pulse–echo scanners a constant velocity of 1540 m/sec is assumed for all tissues, ultrasound velocity in tissues may vary to the extent that wavelets arriving at a point in space have significantly different phase, or phase aberration may occur. This is especially significant for phased array scanners, which use delay lines to focus and steer the beam. Two approaches among others have been used to correct for phase aberration due to the intervening tissues such as body wall between the probe and the region of interest. One uses speckle brightness as a factor for optimization (Nock *et al.*, 1989) and one uses the cross-correlation functions between neighboring elements of a phased array for optimizing the image (O'Donnell and Flax, 1988). In the former, time delays of a phased array are modified to maximize the mean speckle brightness in a region of interest and in the latter, optimal time delays are derived from the phases of cross-correlation functions of adjacent channels computed from the RF A lines from the tissue. Both approaches have been shown able to improve the image quality at the sacrifice of imaging time.

# VI.  Biological Effects of Ultrasound

## A.  Acoustic Phenomena at High Intensity Levels

At high ultrasonic intensities, a number of nonlinear effects appear. So do new acoustic phenomena. Among the most important are wave distortion and cavitation (NCRP, 1983).

### 1.  Wave Distortion

As the power level of the acoustic wave is increased, the sinusoidal pressure wave is distorted. This is because when the medium is compressed, the propagation velocity, inversely proportional to the compressibility and the density, increases.

Thus, in regions of increased pressure, the propagation velocity is greater, causing the pressure peaks to catch up with the pressure troughs. When this occurs, the sinusoidal waveform begins to look like a sawtooth waveform and significant energy is transferred to higher harmonics of the wave, resulting in a higher absorption.

## 2.  Cavitation

The term cavitation is used to describe the behavior of gas bubbles in ultrasonic fields. Two different types of cavitation may occur: transient or stable cavitation.

**Transient Cavitation**    Transient cavitation is the phenomenon in which microbubbles suddenly grow and collapse in a liquid medium. The physical process can be described as follows: Bubbles in a medium are greatly expanded when pressure decreases rapidly. The pressure increases one-half cycle later causing bubbles to collapse and disappear. As shown in Fig. 129, for a very large pressure swing the radius increases markedly, reaching a peak well past the pressure minimum, and as the pressure reaches a peak, the bubble collapses. The internal bubble pressure can become very high, up to 80,000 atm, with a temperature approaching 10,000 K. Such high temperature can cause decomposition of water into chemically active acidic components, causing serious biological effects.

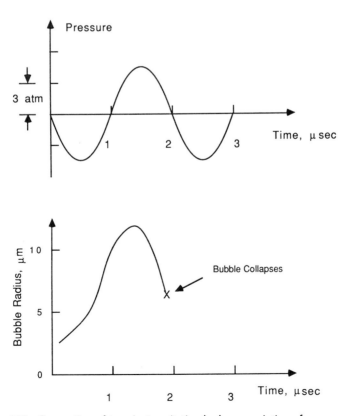

**Figure 129**  Generation of transient cavitation by large variation of pressure.

**Stable Cavitation**   For lower ultrasonic intensities, the bubbles will not collapse as just described. Under this circumstance, the behavior of such bubbles is quite stable and is known as stable cavitation.

Cavitation can be suppressed by degassing the liquid or by increasing the external pressure applied to the system. As can be seen from Fig. 129, it takes a finite amount of time for the gas bubble to respond to the pressure change. Therefore, if the frequency is high enough, the cavitation phenomenon should disappear.

## B.   Ultrasound Bioeffects

Ultrasound bioeffects can be classified into thermal effects or nonthermal effects such as those caused by cavitation. However, in reality on animal preparations these effects are usually difficult to separate. A large body of data has been accumulated over the years from both water tank studies and animal studies in an effort to establish whether diagnostic ultrasound produces any biological effects. It is certain that ultrasound peak intensity higher than 1 watt/cm$^2$ applied continuously for an extended period of time, say, a few minutes, can result in some forms of biological effects, such as fetal weight reduction or increase in the fatality rate of the litters if pregnant animals are irradiated. Unfortunately, even to date experimental results on the bioeffects of ultrasound intensity produced by diagnostic scanners, typically with a spatial peak temporal average intensity in the order of 10 to 100 mW/cm$^2$ (Doppler devices typically have higher power output), are inconclusive. Bioeffects observed in one laboratory sometimes cannot be duplicated at another laboratory using the same equipment and same experimental conditions. However, in surveying all the data that have been accumulated so far it is safe to say that at present in the lower megahertz range (1) no substantial bioeffects have been demonstrated for SPTA intensities less than 100 mW/cm$^2$ and (2) no substantial bioeffects have been demonstrated at even higher intensities for which the product of SPTA intensity and exposure time is less than 50 J/cm$^2$ for exposure time between 1 and 500 sec (AIUM, 1984).

When bioeffects are monitored in suspension under *in vitro* conditions, and temperature elevation in general is not a factor; the mechanism involved is usually some form of cavitation activity. It is often found that cavitation does not occur in an ultrasonic field unless a certain threshold intensity is exceeded. Typically, the value of this threshold depends on the state of nucleation, that is, on the number, size, and distribution of small gaseous bodies. Since nucleation often depends on accidental circumstances, such as the presence of gas trapping impurity particles or the existence of small cracks or crevices in a container, it is understandable that the threshold for cavitation varies greatly from one circumstance to another. Also, it is not surprising that the threshold for cavitation-dependent bioeffects is variable. The threshold for cavitation ranges from a few watts/cm$^2$ to a few hundred watts/cm$^2$, depending on the experimental conditions.

In summary, it seems that SPTA ultrasound intensity less than 100 mW/cm$^2$ should be considered safe, based on currently available experimental information. Commercial diagnostic ultrasonic instruments operated at an intensity lower than 100 mW/cm$^2$ should be considered noninvasive at the present stage of knowledge.

# Problems

1. An ultrasonic wave of 3 MHz with an intensity of 10 mW/cm$^2$ is incident on a flat boundary normally between two media with acoustic impedance $Z_1$ and $Z_2$. Assume that the attenuation in the two media can be neglected. The reflected power received by an ultrasonic transducer with an aperture area of 0.1 cm$^2$ is 0.1 mW. **(a)** What is the transmitted intensity? **(b)** If $Z_1$ is assumed to be $1.5 \times 10^6$ kg/m$^2$-sec, find $Z_2$. **(c)** Find reflected and transmitted pressure and show that $P_i + P_r = P_t$. **(d)** Could $P_t$ possibly be greater than $P_i$? Why?

2. Calculate the distance a 1-MHz ultrasound will travel before its intensity drops to one-half its incident value for (a) air, (b) aluminum, (c) blood, (d) liver. Repeat for 5-MHz ultrasound.

3. It has been shown that the scattering of ultrasound by cardiac muscle is approximately proportional to the third power of the frequency. Explain why the attenuation of ultrasound in cardiac muscle is linearly dependent on frequency.

4. An investigator wants to measure the ultrasonic attenuation coefficients of a number of materials including liquids and solids. He sets up the following experimental scheme. He thinks that the attenuation coefficient of a liquid can be measured by moving the receiving transducer a predetermined distance and then determining the change in amplitude of the received signal. Suppose that two transducers of 1 cm diameter with a resonant frequency of 5 MHz are used. To calibrate the system, he uses a liquid of known attenuation coefficient, say, blood. He finds that the amplitude of the signal drops much more than anticipated after he moves the transducer backward by 2 cm. **(a)** What is the reason for this discrepency? He then inserts a slab of aluminum with a thickness of 3 cm into the beam as shown. **(b)** Should he observe an increase or a decrease in the received intensity? Explain.

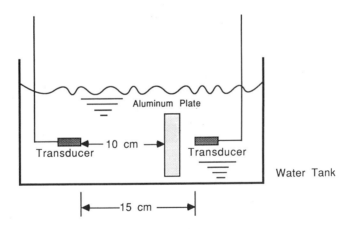

5. The spatial beam profile of an ultrasonic transducer 5 cm away from the transducer in water can be approximated as shown in (a). The intensity waveform at the location $x = 0$ has been determined and is shown in (b). What are the I(SATA), I(SPTA), I(SPTP), and I(SATP)?

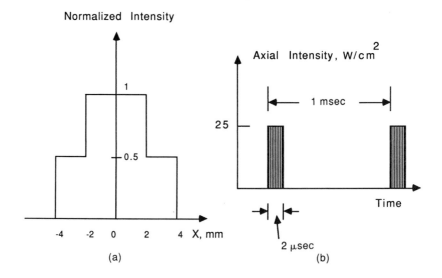

(a)    (b)

6. A very precise technique for measuring ultrasonic attenuation in a liquid is illustrated in the accompanying figure. The water and the sample liquid are separated by a thin membrane whose attenuation is negligible. Two transducers are mounted on a movable carriage with a fixed distance $d$ between them. Suppose the attenuation coefficient of water is given by $\beta_0$. The attenuation coefficient of the sample, $\beta$, can be readily calculated by observing the change in received pressure after a displacement of carriage, say, to the right, by $\Delta d$. Derive an expression for $\beta$ in terms of known parameters, $\beta_0$, $\Delta d$, $d$, and the measured pressures before and after carriage displacement.

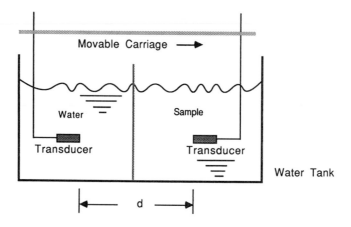

7. A PZT-4 disc is used to generate a 5-MHz ultrasonic beam with the narrowest beam region centered about 5 cm from the transducer in water. Find the thickness and diameter of the disc.

8. A clinician is using a linear array to image a liver. Suppose that in this liver there is a large tumor shaped as shown containing a high concentration of collagen whose acoustic speed is much larger than the liver tissue. Draw an image of the structure. Would this cause any distortion to the echo of the back surface of the liver? If your answer is yes, explain.

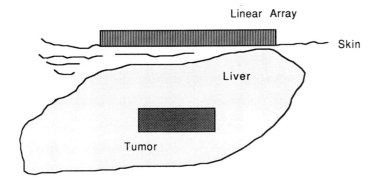

9. A real-time linear array is used to image an interface as shown. Eight elements are fired as a group. The radiation pattern of the 8-element group is also shown. Assume that the interface is in the far field of the transducer. Draw a B-scan image of the interface to illustrate the artifacts produced by the grating lobes of the array.

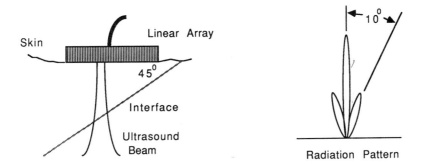

10. A 128-element linear array is used to image the uterus in real time. A full-term uterus may have a thickness of 50 cm. Suppose that groups of four elements are fired sequentially. **(a)** How many scan lines are there? **(b)** What is the frame rate if sound speed in the uterus is assumed to be $1.5 \times 10^5$ cm/sec? **(c)** What can you do to raise the frame rate?

11. A three element annular array 1.8 cm in diameter as shown is used to image the heart. The average sound speed in heart tissues is $1.6 \times 10^5$ cm/sec. Calculate the time delays of the electrical pulses feeding the annular elements relative to the outmost elements if the focal point is required to be 10 cm away from the array.

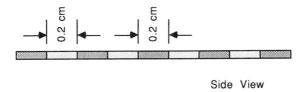

Side View

12. An 8-MHz pulsed Doppler flowmeter is used to monitor blood flow in a carotid artery (Fig. 116). The probe is making a 45° angle with respect to the direction of flow. Suppose the velocity of the blood flow in the artery is uniform throughout the lumen and is approximately 50 cm/sec. The distance between the probe and the

back arterial wall is 3 cm. Estimate the suitable range for the pulse repetition frequency of the device. Average sound velocity in tissue $= 1.55 \times 10^5$ cm/sec.

13. A pulse of ultrasound $i(t)$ is incident on a slab of biological tissue as shown. The Fourier transform of $i(t)$ is $I(f)$. The Fourier transform of the transmitted pulse denoted as $r(t)$, $R(f)$ is related to $I(f)$ by the following relationship from system theory:

$$R(f) = H(f)I(f)$$

where $H(f)$ is the transfer function representing the frequency characteristic of the

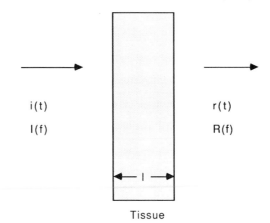

Tissue

slab and is the Fourier transform of the impulse response, $h(t)$. **(a)** Obtain $H(f)$ for the arrangement shown and $h(t)$ assuming the attenuation coefficient is linearly dependent on frequency, that is, $\beta(f) = \beta_0 f$ where $\beta_0$ is the attenuation coefficient of the tissue at 1 MHz. **(b)** Show that $\beta$ can be estimated from the root mean square (RMS) duration $D_t$ of the impulse response. $D_t$ is defined as

$$D_t = \left[ \frac{\int_{-\infty}^{\infty} (t - t_c)^2 |h(t)|^2 \, dt}{\int_{-\infty}^{\infty} |h(t)|^2 \, dt} \right]^{1/2}$$

where $t_c = 1/c$, l is the thickness of the slab, and $c$ is the sound velocity in tissue.

14. In designing ultrasonic transducers, sometimes a quarter-wavelength layer is used to minimize acoustic impedance mismatch between the transducer and the medium in front of the transducer. This is because total transmission occurs if $l_2 = \lambda/4$ and $Z_2 = (Z_1 Z_3)^{1/2}$. **(a)** Obtain the expression for $I_t/I_i$. **(b)** Show $I_t/I_i = 1$ if $Z_2 = (Z_1 Z_3)^{1/2}$ and $l_2 = \lambda/4$.

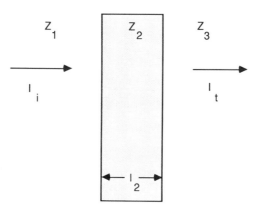

# References and Further Reading

American Institute of Ultrasound in Medicine (1984). "Safety Considerations for Diagnostic Ultrasound." AIUM, Bethesda, Maryland.

Atkinson, P., and Woodcock, J. P. (1982). "Doppler Ultrasound and Its Use in Clinical Measurements." Academic Press, London and New York.

Bolt, R. H., and Hueter, T. F. (1955). "Sonics." John Wiley, New York.

Bonnefous, O., and Pesgue, P. (1986). *Ultrasonic Imaging* **8**, 73.

Brekhovskikh, L. M. (1960). "Waves in Layered Media." Academic Press, New York.

Busse, L. J., and Miller, J. G. (1981). *J. Acoust. Soc. Am.* **70**, 1377.

Carstensen, E. L. (1979). Absorption of sound in tissues *In* "Ultrasonic Tissue Characterization II." National Bureau of Standards special publication 524, Gaithersberg, Maryland.

Chen, D. E., Robinson, L. S., Wilson, K. A., Griffiths, A., and Doust, B. D. (1987). *Ultrasonic Imaging* **9**, 221.

Christensen, D. A. (1988). "Ultrasonic Instrumentation." John Wiley, New York.

Cobb, W. N. (1983). *J. Acoust. Soc. Am.* **73**, 1525.

Dainty, J. C. (1984). "Laser Speckle." Springer-Verlag, Berlin and New York.

deBruign, N. P., and Clements, F. M. (1987). "Transesophageal Echocardiography." Martinus Nijhoff, Boston.

DeSilets, C. S., Fraser, J. D., and Kino, G. S. (1978). *IEEE Trans. Sonics Ultrasonics* **25**, 115.

Durnin, J. (1987). *J. Opt. Soc. Am.* **4**, 651.

Embree, P. M., and O'Brien, W. D., Jr. (1990). *IEEE Trans. Ultras. Ferroelec. and Freq. Cont.* **37**, 176.

Evans, D. H., McDicken, W. N., Skidmore, R., and Woodcock, J. P. (1989). "Doppler Ultrasound: Physics, Instrumentation, and Clinical Applications." John Wiley, New York.

Fei, D. Y., and Shung, K. K. (1985). *J. Acoust. Soc. Am.* **78**, 871.

Feinstein, S. B., Cheirif, J., Ten Cate, F. J., Silverman, P. R., Heidenreich, P. A., Dick, C., Desir, R. M., Armstrong, W. F., Quinones, M. A., and Shah, P.(1990). *J. Am. Coll. Cardiol.* **16**, 316.

Feleppa, E. J., Lizzi, F. L., Coleman, D. J., and Yaremko, M. M. (1986). *Ultras. Med. Biol.* **14**, 462.

Fields, S., and Dunn, F. (1973). *J. Acoust. Soc. Am.* **54**, 809.

Flax, S. W., Pelc, N. J., Glover, G. H., Gutmann, F. D., and McLachlan, M. (1981). *Ultrasonic Imaging* **5**, 95.

Geleskie, J. V., and Shung, K. K. (1982). *J. Acoust. Soc. Am.* **71**, 467.

Goss, S. A., Johnson, R. L., and Dunn, F. (1978). *J. Acoust. Soc. Am.* **64**, 423.

Goss, S. A., and O'Brien, W. D., Jr. (1979). *J. Acoust. Soc. Am.* **65**, 507.

Goss, S. A. and O'Brien, W. D., Jr. (1980). *J. Acoust. Soc. Am.* **67**, 1041.

Goldberg, B. B., and Kimmelman, B. A. (1988). "Medical Diagnostic Ultrasound: A Retrospective on Its 40th Anniversary." Kodak, Rochester, New York.

Greenleaf, J. F. (1983). *IEEE Proc.* **71**, 330.

Greenleaf, J. F. (1986). "Tissue Characterization with Ultrasound." CRC Press, Boca Raton, Florida.

Gururaja, T. R., Schultz, W. A., Cross, L. E., and Newham, R. E. (1985). *IEEE Trans. Sonics Ultrasonics* **32**, 499.

Hatle, L., and Angelsen, B. (1985). "Doppler Ultrasound in Cardiology, 2nd ed." Lea and Febiger, Philadelphia.

Heyman, J. S. (1978). *J. Acoust. Soc. Am.* **64**, 243.

Insana, M. F., and Hall, T. J. (1990). *Ultrasonic Imaging* **12**, 245.

Ishimaru, A. (1978). "Wave Propagation and Scattering in Random Media." Academic Press, San Diego, California.

Kasai, C., Namekawa, J., Koyano, A., and Omoto, R. (1985). *IEEE Trans. Sonics Ultrasonics* **32**, 458.

Kessler, L. W. (1985). *IEEE Trans. Sonics Ultrasonics* **32**, 136.

Kino, G. S. (1987). "Acoustic Waves: Devices, Imaging, and Analog Signal Processing." Prentice-Hall, Englewood Cliffs, New Jersey.

Kinsler, L. E., and Frey, A. R. (1962). "Fundementals of Acoustics, 2nd ed." John Wiley, New York.

Kisslo, J., Adams, D. B., and Belkin, R. N. (1988). "Doppler Color Flow Imaging." Churchill Livingstone, Edinburg, U.K.

Kuc, R. (1984). *IEEE Acoust. Speech Sig. Proc. Mag* **1**, 19.

Larson, J. S. III (1983). *Hewlett-Packard J.* **34**, 17.

Law, W. K., Frizzell, L. A., and Dunn, F. (1983). *J. Acoust. Soc. Am.* **74**, 1295.

Lewin, P. A. (1981). *Ultrasonics* **19**, 213.

Linzer, M. (1976). "Ultrasonic Tissue Charaterization I." National Bureau of Standards Special Public. 453, Gaithersberg, Maryland.

Linzer, M. (1979). "Ultrasonic Tissue Charaterization II." National Bureau of Standards Special Public. 524, Gaithersberg, Maryland.

Lockwood, G. R., Ryan, L. K., Hunt, J. W. and Foster, F. S. (1991). *Ultras. Med. Biol.* **17**, 653.

Lu, J. L., and Greenleaf, J. F. (1990). *IEEE Trans. Ultras. Ferro. Freq. Cont.* **37**, 438.

Madsen, E. L. (1986). Ultrasonically soft tissue-mimicking materials and phantoms. *In* "Tissue Characterization with Ultrasound". (J. F. Greenleaf ed.), CRC Press, Boca Raton, Florida.

Macovski, A. (1979). *IEEE Proc.* **67**, 484.

Magnin, P. A., Von Ramm, O. T., and Thurstone, F. L. (1982). *Ultrasonic Imaging* **4**, 267.

Malecki, I. (1969). "Physical Foundations of Technical Acoustics." Pergamon Press, New York.

Marcus, P. W. and Carstensen, E. L. (1975). *J. Acoust. Soc. Am.* **58**, 1334.

McCann, H. A., Sharp, J. C., Kinter, T. M., McEwan, C. N., and Greenleaf, J. F. (1988). *IEEE Proc.* **76**, 1063.

McDicken, W. N. (1991). "Diagnostic Ultrasonics: Principles and Use of Instruments 3rd Ed." Churchill Livingstone, Edinburgh, U.K.

Meltzer, R. S., and Roelandt, J. (1982). "Contrast Echocardiography." Martinns Nijhoff, Boston.

Miller, J. G., Perez, J. E., and Sobel, B. E. (1985). *Prog. Cardiov. Dis.* **28**, 85.

Mulunski, M. R., Mohr, G. A., Perez, J. E., Vered, Z., Wear, K. A., Gessler, C. J., Sobel, B. E., Miller, J. G., and Wickline, S. A. (1980). *Circulation* **80**, 491.

Morse, P. M., and Ingard, K. U. (1968). "Theoretical Acoustics." McGraw-Hill, New York.

National Council on Radiation Protection and Measurements (1983). "Biological Effects of Ultrasound: Mechanisms and Clinical Implications." NCRP, Bethesda, Maryland.

Nock, L., and Trahey G. E. (1989). *J. Acoust. Soc. Am.* **85**, 1819.

O'Donnell, M., Mimbs, J. W., and Miller, J. G. (1979). *J. Acoust. Soc. Am.* **65**, 512.

O'Donnell, M., Mimbs, J. W., and Miller, J. G. (1981). *J. Acoust. Soc. Am.* **69**, 580.

O'Donnell, M., and Flax, S. W. (1988). *IEEE Trans Ultras. Ferro. Freq. Cont.* **35**, 768.

Ophir, J., Cespedes, I., Ponnekanti, H., Yazdi, Y., and Li, X. (1991). *Ultrasonic Imaging* **13**, 111.

Pauly, H., and Schawn, H. P. (1970). *J. Acoust. Soc. Am.* **50**, 692.

Parker, K. J. (1983). *Ultras. Med. Biol.* **9**, 363.

Parker, K. J., Tuthill, T. A., Lerner, R. M., and Violante, M. R. (1987). *Ultras. Med. Biol.* **13**, 555.

Parker, K. J., Huang, S. R., Musulin, R. A., and Lerner, R. M. (1990). *Ultras. Med. Biol.* **16**, 241.

Rose, J. L., and Goldberg, B. B. (1981). "Basic Physics in Diagnostic Ultrasound." John Wiley, New York.

Schwan, H. P. (1970). "Biological Engineering." John Wiley, New York.

Sessler, G. M. (1981). *J. Acoust. Soc. Am.* **70,** 1596.

Shapiro, J. R., Reisner, S. A., and Meltzer, R. S. (1989). *J. Am. Coll. Cardiol.* **13,** 1629.

Shattuck, D. P., Weinshenker, M. D., Smith, S. W., and von Ramm, O. T. (1984). *J. Acoust. Soc. Am.* **75,** 1273.

Shung, K. K., and Reid, J. M. (1977). *IEEE 1977 Ultras. Sym. Proc.* 230.

Shung, K. K. (1987). *CRC Crit. Rev. Biomed.* Eng. **15,** 1.

Smith, S. W., Pavy, H. E., Jr., von Ramm, O. T. (1991). *IEEE Trans. Ultras. Ferro. Freq. Cont.* **38,** 100.

Smith, W. A. (1989). *IEEE 1985 Ultras. Sym. Proc.* 755.

Trahey, G. E., Hubbard, S. M., and von Ramm, O. T. (1988). *Ultrasonics* **27,** 271.

Trahey, G. E., Smith, S. W., and von Ramm, O. T. (1986). *IEEE Trans. Ultras. Ferro. Freq. Cont.* **33,** 257.

Twersky, V. (1978). *J. Acoust. Soc. Am.* **64,** 1710.

von Ramm, O. T., Smith, S. W., and Pavy, H. E. (1991). *IEEE Trans. Ultras. Ferro. Freq. Cont.* **38,** 109.

Wade, G. (1976). "Acoustic Imaging." Plenum Press, New York.

Wagner, R. F., Smith, S. W., Sandrick, J. M., and Lopez, H. (1983). *IEEE Trans Sonics Ulrasonics* **30,** 156.

Wells, P. N. T. (1977). "Biomedical Ultrasonics." Academic Press, New York.

Yock, P. G., Johnson, E. L., and Linker, D. T. (1988). *Am. J. Cardiac. Im.* **2,** 185.

Yuan, Y. W., and Shung, K. K. (1988). *J. Acoust. Soc. Am.* **84,** 52.

# CHAPTER 3

# Radionuclide Imaging

Radioactivity from naturally occurring radioisotopes was discovered by Becquerel in 1896. In 1934, the Curies (husband Jean Frederic and wife Irene) produced radiophosphorous $^{32}P$, the first artificial radioisotope, and coined the words *radioactive* and *radioactivity*. These historical events started nuclear medicine, a branch of medicine that utilizes mainly artificial radioisotopes in the diagnosis and treatment of diseases. With the advance in the production of artificial radionuclides spurred by the development of high-energy cyclotrons and particle accelerators, nuclear medicine has grown into an important clinical and research area in modern medicine.

The concept and the term *radionuclide* were proposed by Kohman in 1947 to describe a nucleus with a measurable half-life of radioactive decay. The use of radionuclides for diagnostic purposes was pioneered by Hevesy, who first used radiophosphorous for metabolic studies in rats in 1935. The Geiger–Muller (GM) tube was used in the initial studies and continued to be utilized in the early human studies. The era of radionuclide imaging was initiated in 1949 when Cassen developed a prototype rectilinear scanner consisting of a calcium tungstate crystal coupled to a photomultiplier tube. By using a lead collimator and a scanning mechanism, he was able to obtain an image of $^{131}I$ uptake in the thyroid with 1/4-inch spatial resolution. Rectilinear scanners become the major equipment for radionuclide imaging in the 1950s and early 1960s, when substantial improvement was made.

Perhaps the most important instrumentation advance made in radionuclide imaging was the introduction of a stationary area detector known as the scintillation camera (also known as Anger or gamma camera) by Anger in 1957. The scintillation camera consists of a large-area sodium iodide crystal coupled to an array of photomultiplier tubes. It allows rapid acquisition of radionuclide images without the mechanical scanning motion. The scintillation camera has become the single most important imaging device for clinical nuclear medicine since the mid-1960s.

The idea of transverse section tomography was introduced by Kuhl and Edwards in the early 1960s, years before the introduction of X-ray CT (computed tomography). They constructed the Mark IV scanner, an imaging system consisting of a square array of multiple collimated detectors that surrounded the patient's head. The system was capable of rotate–translate motion in acquiring projection data from different angular views. However, without the aid of computers and knowledge of reconstruction algorithms from projection data, the tomograms produced by the system were less than satisfactory. Single-photon emission computed tomography (SPECT) as known today was poineered independently by Keyes and Jaszczak in 1977 (the latter author coined the acronym SPECT). It provides substantial improvement in image quality, especially in terms of image contrast, over the conventional radionuclide imaging technique using stationary cameras.

The principle of positron emission tomography (PET) is based on the coincidence detection of the two 511-keV photons emitted in opposite directions following the annihilation of a positron and an electron. Development of PET was at its peak in the 1970s when a number of major PET systems were designed and constructed. A typical PET system consists of a large number of scintillation detectors that surround the patient; each detector is in coincidence with those on the opposite side of the patient. The coincidence detection provides the projection data for reconstruction of positron-emitting radionuclide distribution within the patient. Since almost all of the body constituents have positron-emitting radioisotopes (e.g., $^{11}C$, $^{15}O$, $^{13}N$), PET offers a unique capability to study metabolic functions in humans. However, the positron-emitting radionuclides for PET imaging have short half-lives (e.g., 20 min for $^{11}C$, 2 min for $^{15}O$ and 10 min for $^{13}N$), thus requiring an on-site cyclotron for production. The high costs of PET systems and on-site cyclotrons have limited PET to major clinical and research centers.

Radionuclides can be labeled with various pharmaceuticals to form radiopharmaceuticals. Administration of these radiopharmaceuticals through injection, inhalation, or orally brings the radionuclide to the designated tissues or passage ways. The "trace" amount of radiopharmaceuticals (in the order of nanograms) used in clinical nuclear medicine has little effect on the biochemical and physiological functions of the systems under study. The measurement of biochemical and physiological functions using radionuclides and sensitive imaging techniques are the fundamental basis of nuclear medicine.

Although radionuclide imaging suffers from relatively poor spatial resolution and high image-noise fluctuations, it provides information about the physiological functions of the patient that is difficult or impossible to obtain from other imaging modalities such as X-ray or ultrasound techniques. The rapid advances in radiopharmaceuticals, instrumentation, imaging processing techniques, and clinical applications made in the last decade have greatly enhanced radionuclide imaging as an important modern medical imaging modality.

# I.   Fundamentals of Radioactivity

## A.   Nuclear Particles

A number of naturally occurring unstable or radioactive isotopes emit radiation and/or nuclear particles when they disintegrate. The most well-known types are the alpha, beta, and gamma emissions; their properties are given in Table VII. Nuclear stability has been found to be dependent on the neutron/proton ratio within the nucleus. The importance of the neutron/proton ratio in determining nuclear stability is illustrated in Fig. 130 where the solid line shows the neutron/proton ratio for stable nuclei and the dotted line shows the line for the ratio equal to 1. At low atomic number the nuclei are stable when the numbers of neutrons and protons are equal, whereas at high atomic number, the nuclei are stable when the number of neutrons is approximately 1.5 times the number of protons. Nuclei with an unfavorable neutron/proton ratio will disintegrate or decay into stable nuclei by spontaneous emission of nuclear particles. For example, a neutron-rich nucleus will undergo the conversion represented schematically by

$$n \longrightarrow p^+ + e^- + \bar{\nu}_e + \text{energy} \tag{3-1}$$

where $n$, $p^+$, $e^-$ and $\bar{\nu}_e$ represent neutron, proton, electron or beta particle, and anti-electron neutrino, respectively. On the other hand, a proton-rich nucleus may undergo the conversion

$$p^+ \longrightarrow n + e^+ + \nu_e + \text{energy} \tag{3-2}$$

where $\nu_e$ is an electron neutrino and $e^+$ is the antiparticle of an electron, also known as positron. A positron has the same mass as an electron but carries a positive charge of the same magnitude as an electron. In the above two equations, the total rest masses of all the particles on either side are different. Additional energy results to balance these equations.

An element with a given atomic number has a fixed number of protons in its nucleus (e.g., carbon has six protons in its nucleus). However, the number of neutrons in the nucleus may vary. Elements with the same number of protons but a different number of neutrons are called isotopes. For examples, carbon has two stable isotopes, $^{12}C$ and $^{13}C$, and a naturally occurring radioisotope, $^{14}C$. Isotopes can also be produced artificially by bombarding a nucleus with high-energy particles (e.g., $^{11}C$ and $^{15}C$).

All isotopes used in nuclear medicine are artificially produced and radioactive. The radioactive isotopes are also called radioisotopes or radionuclides. Table VIII lists some of the characteristics of important radionuclides used in nuclear

**Table VII**
Characteristics of particles emitted by decaying unstable nuclei

| Particle | Rest mass (kg) | Charge (electron charge) | Nature | Energy distribution |
|---|---|---|---|---|
| Alpha | $6.6 \times 10^{-27}$ | +2 | helium nucleus | monoenergetic |
| Beta | $9.0 \times 10^{-31}$ | −1 | electron | continuous |
| Gamma | 0 | 0 | photon | monoenergetic |

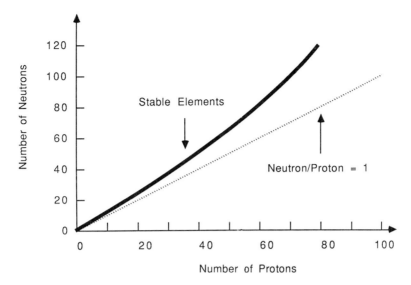

**Figure 130**  All stable elements have the neutron/proton ratio indicated by the bold line.

medicine. Among these radionuclides, $^{99m}$Tc is the single most popular one used in clinical practice. The gamma radiation from these radionuclides is primarily the emission used in imaging due to its high penetrating power through thick tissues compared to the alpha and beta emissions.

Radioactive decay is the process by which an unstable nucleus is transformed into a more stable daughter nucleus. Depending on the specific mode of decay, the transformation often involves particle and/or photon emissions and the release of nuclear energy. Table IX lists the various modes of radioactive decay and the accompanying emissions (A = atomic mass number; Z = atomic number).

For example, in electron capture, a proton-rich nucleus may transform to the stable state by capturing one of its own orbital electrons. The electron combines with a proton to form a neutron and a neutrino. The transition energy is released as kinetic energy of the neutrino. Approximately 90% of electron capturing in-

**Table VIII**
Characteristics of important radionuclides

| Element | Radionuclide | Emission | Photon energy (MeV) | Half-life |
|---|---|---|---|---|
| Carbon | $^{11}$C | $e^+$ | 0.511 | 20 min |
| Nitrogen | $^{13}$N | $e^+$ | 0.511 | 10 min |
| Oxygen | $^{15}$O | $e^+$ | 0.511 | 124 sec |
| Fluorine | $^{18}$F | $e^+$ | 0.511 | 109 min |
| Cobalt | $^{57}$Co | $\gamma$ | 0.122, 0.136 | 270 days |
| Gallium | $^{68}$Ga | $\gamma$ | 0.511, 1.077 | 68 min |
| Technetium | $^{99m}$Tc | $\gamma$ | 0.14 | 6 hrs |
| Indium | $^{113m}$In | $\gamma$ | 0.393 | 102 min |
| Iodine | $^{123}$I | $\gamma$ | 0.159 | 13 hrs |
| Iodine | $^{131}$I | $\gamma$ | 0.080, 0.284, 0.364 | 8 days |
| Thallium | $^{201}$Tl | $\gamma$ | 0.135, 0.167 | 73 hrs |

**Table IX**
Modes of radioactive decay

| Radioactive decay | Parent nucleus | Daughter nucleus | Emission |
|---|---|---|---|
| Alpha decay | $_Z^A X$ | $_{Z-2}^{A-4} Y$ | $\alpha$, $\gamma$ |
| Beta decay | $_Z^A X$ | $_{Z+1}^A Y$ | $e^-$ (continuous energy), $\nu$, $\gamma$ |
| Positron decay | $_Z^A X$ | $_{Z-1}^A Y$ | $e^+$ (continuous energy), $\nu$, $\gamma$ |
| Isomeric decay | $^m X$ | $X$ | $\gamma$ |
| Electron capture | $_Z^A X + e_k^-$ | $_{Z-1}^A Y$ | $\nu$, $\gamma$, characteristic X-rays, Auger electrons |
| Internal conversion | $^m X$ | $X$ | $e_k^-$ (ejected), characteristic X-rays, Auger electrons |

volves K-shell electrons, which are closest to the nucleus. There could be gamma radiation associated with electron capture. When the electron vacancy is filled, additional energy is released in the form of characteristic X-rays and Auger electrons.

In isomeric decay, a metastable radionuclide decays by emitting $\gamma$ rays, and the daughter nucleus differs from its parent only in that it is in its stable state. An important example is $^{99m}$Tc, which decays by emitting gamma photons with 140keV energy. An alternative mode of decay by nucleus in metastable state is internal conversion where the energy of the nucleus is released by ejecting an orbital electron. The ejected electron is called a conversion electron. The orbital electron vacancy is subsequently filled by an outer electron with the emission of characteristic X-rays and Auger electrons.

Another important decay in medical imaging is the positron decay. When a positron collides with an electron, both particles annihilate to form two 511-keV photons moving in opposite directions. The use of radionuclides with positron decay and the detection of annihilation photons form the basic principles of positron emission tomography (PET).

## B.   Nuclear Activity and Half-Life

Radioactivity decay can be described by the mathematical expression

$$N(t) = N_0 e^{-\lambda t} \tag{3-3}$$

where $N_0$ and $N(t)$ are the numbers of radionuclide at time $t = 0$ and $t$, respectively. The factor $e^{-\lambda t}$ is the fraction of radionuclide remaining after time $t$ and is called the decay factor. The decay constant $\lambda$ is related to the half-life of the radionuclide $T_{1/2}$, defined as the time required for half of the radionuclides to decay by

$$T_{1/2} = \frac{0.693}{\lambda} \tag{3-4}$$

Figure 131 shows the plot of decay factor as a function of time for $^{99m}$Tc.

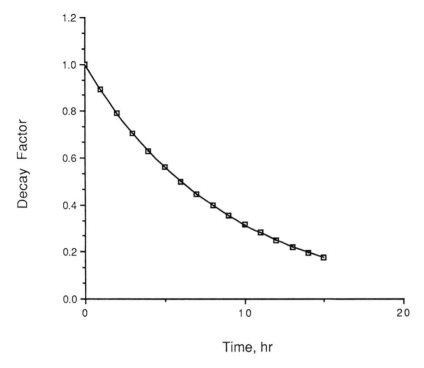

**Figure 131**   Decay factor of $^{99m}$Tc as function of time.

The activity $A$ of a radionuclide is defined as the average decay rate, or the time derivative of Eq. (3-3), that is,

$$A(t) = -\frac{dN(t)}{dt} = \lambda N(t) \tag{3-5}$$

The dimension of activity $A$ can be expressed in terms of decays per second (dps) or decays per minute (dpm). Since $A(t)$ is linearly proportional to $N(t)$, it follows that (where $A_0 = \lambda N_0$)

$$A(t) = A_0 e^{-\lambda t} \tag{3-6}$$

## C.   Units for Measuring Nuclear Activity

Two common units have been used as measures of activity, the curie and the becquerel. A curie (Ci) is defined as $3.7 \times 10^{10}$ disintegrations or decays per second (dps). It was defined originally as the activity of one gram of $^{226}$Ra. In nuclear medicine, a curie represents a very large amount of activity. The units millicurie (1 mCi = $10^{-3}$ Ci) or microcurie (1 $\mu$Ci = $10^{-3}$ mCi = $10^{-6}$ Ci) are often used. The SI (Système International) unit of activity is the becquerel (Bq), defined as one decay per second. Thus, 1 Ci = $3.7 \times 10^{10}$ Bq.

## D.   Specific Activity

Specific activity is defined as the ratio of activity to the mass of the radionuclide sample. The reason for introducing this additional parameter is that a radionuclide sample may contain stable isotopes of the element, which are called the carrier. The specific activity of a radionuclide sample is highest when it is carrier-free. The carrier-free specific activity of a radionuclide sample is given by Sorenson and Phelps (1987).

$$A_{cf} = \frac{1.3 \times 10^8}{A_m T_{1/2}} \tag{3-7}$$

where $A_{cf}$ in Ci/gm is the carrier-free specific activity, $A_m$ is the mass number, and $T_{1/2}$ is the half-life of the radionuclide in days. A high specific activity is required for tracer study in biomedical applications because a smaller dose of tracer yields more nuclear activity. The specific activities of $^{131}$I and $^{99m}$Tc are $1.2 \times 10^5$ and $5.3 \times 10^6$ Ci/gm, respectively.

## E.   Interaction of Nuclear Particles and Matter

### 1.   Alpha Particles

An alpha particle is a helium nucleus consisting of two neutrons and two positively charged protons. Most radionuclide that decay by alpha emission have large nuclei (in fact most have an atomic number greater than 82). The average energy of an emitted alpha particle ranges from 3 to 9 MeV. The range of an alpha particle in air before being stopped is proportional to its energy raised to the three-half power. Due to its dual positive charge and a large mass, the alpha particle has strong interactive power and loses its energy primarily by the ionization or excitation of other particles or atoms. The ionization ability of an alpha particle can be expressed in terms of the specific ionization defined as the number of ions produced per unit length. The specific ionization for several particles is tabulated in Table X (Rollo, 1977).

The mean range in air, $R_m$, in centimeters for alpha particles with energy $E_a$ in mega-electron-volts can be approximated by

$$R_m = 0.325(E_a)^{3/2} \tag{3-8}$$

The mean range of alpha particles in a medium other than air depends on the density of the medium and can be expressed by

$$R_{mm} = \frac{\eta_a A_{mm}^{1/2}}{\eta_m A_{ma}^{1/2}} \tag{3-9}$$

## Table X
Mean range and specific ionization of various nuclear particles and photons

| Radiation type | Energy | Mean range | | Relative specific ionization |
| --- | --- | --- | --- | --- |
| | | Air | Water | |
| $\alpha$ | 9 MeV | 9 cm | 45 $\mu$m | 2500 |
| $\beta$ | 3 MeV | 10 m | 1 mm | 100 |
| $e^+$ | 3 MeV | 10 m | 1 mm | 100 |
| neutron | 10 MeV | 100 m | 1 m | 0.1 |
| X-ray | 200 keV | 20 m | 1 cm | 10 |
| $\gamma$ | 10 MeV | 100 m | 10 cm | 1 |

Source: Rollo, 1977

where $\eta_a$ and $\eta_m$ are densities of air and of the medium, $R_m$ and $R_{mm}$ are the mean ranges in air and in the medium, and $A_{ma}$ and $A_{mm}$ represent the mass numbers of air and of the medium, respectively.

## 2. Beta Particles

The interaction of beta particles and matter was discussed in Chapter 1 on X-ray generators in which X-rays are produced by striking an anode material with high-energy electrons. In essence, when beta particles interact with a medium, both characteristic and white X-ray radiation will be produced, as shown in Fig. 8.

The range of a beta particle is difficult to determined because of its smaller mass and charge. However, it is found to be proportional to its energy but inversely related to the density of the medium. By using this relationship, one can show that the range of beta particles in different media when normalized to the density of the medium with a unit of gm/cm$^2$, are very similar.

## 3. Gamma Rays

Gamma rays are identical to X-rays except for their origin. X-rays are extranuclear in origin whereas gamma rays are intranuclear. The interaction of low-energy gamma rays and matter is identical to the interaction between X-rays and matter as discussed in Chapter 1. Two additional types of interactions can also occur with gamma rays carrying energy greater than 1.02 MeV. They are pair production and photo disintegration. In pair production, 1.02 MeV out of the total photon energy is converted into the rest mass of an electron–positron pair when the gamma ray photon travels near a large electric field surrounding the nucleus. Photon energy in excess of 1.02 MeV is converted to the kinetic energy of the electron and positron. The interaction is illustrated in Fig. 132. Annihilation occurs when the positron is combined with an electron, resulting in two photons each with 511-keV energy and emitting in opposite directions. In photo disintegration, the gamma ray photon is absorbed by the nucleus, and a neutron, a proton, or an alpha particle is subsequently ejected from the nucleus.

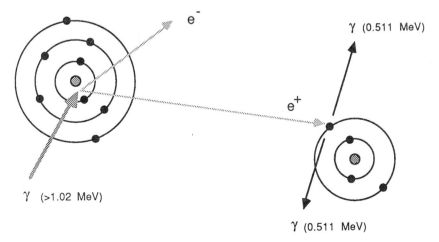

**Figure 132** Pair production as a result of interaction between gamma ray and nucleus.

## F.  Attenuation of Gamma Radiation

The attenuation of gamma rays through a medium is governed by the same equation for X-ray

$$I = I_0 e^{-\beta x} \tag{3-10}$$

where $I_0$ and $I$ are the incident and transmitted intensities of the gamma rays, respectively, $x$ is the thickness of the medium, and $\beta$ is the linear attenuation coefficient. As in X-ray, $\beta$ is a function of the density of the medium and the energy of the gamma photon. A mass attenuation coefficient can be defined similarly. The mass attenuation coefficients for water, sodium iodide, and lead are plotted as functions of photon energy in Fig. 133.

## G.  Radionuclides

Ideal radionuclides for medical diagnosis should possess the following properties.

### 1.  Physical Properties

**Physical Half-Life**    The physical half-life of an ideal radionuclide should be sufficiently short that a substantial amount of radioactivity has been decayed before being excreted by the body. At the same time, it should be long enough for imaging purposes and for transporting from the generator to the patient before the nuclear activity drops to an unacceptably low level. The 6-hour half-life for $^{99m}$Tc is almost ideal for these criteria, which is one of the main reasons for its popularity in nuclear medicine.

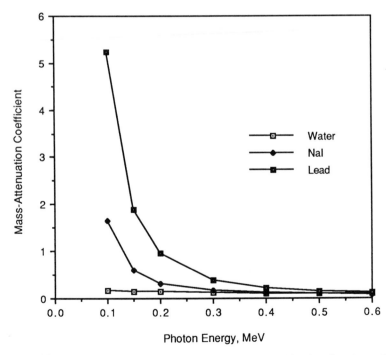

**Figure 133**  Mass attenuation coefficients of water, NaI, and lead as functions of gamma ray energy.

**Types of Emissions**   Since minimizing radiation dose to the patient is a major concern in radionuclide imaging, charged particle emissions are undesirable as they produce ionization, which is the main cause of radiation dose in the patient. Radiation detectors used in radionuclide imaging are designed for detecting gamma radiation. For optimum imaging, the energy of the gamma ray photons should be high enough that most of the photons will escape from the patient. At the same time, the energy should be low enough for detection by the scintillation crystal and without substantial penetration through the collimators. The 140-keV gamma-ray photons, the major emission from $^{99m}$Tc, are almost ideal for radiation dose and imaging considerations.

**Specific Activity**   As described earlier, radionuclide samples with high specific activity are desirable for tracer studies using imaging methods.

**Preparation Method**   It is preferred that the radionuclides can be produced easily with high yield. For those that are generated by accelerators, it is desirable that the nuclear reaction has high cross section. Radionuclide that can be produced by portable generator such as $^{99m}$Tc are especially desirable.

## 2.   Biological Properties

**Biological Half-Life**   The time needed for the body to excrete half of the amount of radionuclide administered is defined as its biological half-life. If the excretion process can be described by an exponential decay, the effective half-life of the radionuclide, $T_{1/2}^{\text{eff}}$, is given by

$$T_{1/2}^{\text{eff}} = \frac{T_{1/2}^{\text{p}} T_{1/2}^{\text{b}}}{(T_{1/2}^{\text{b}} + T_{1/2}^{\text{b}})} \tag{3-11}$$

where $T_{1/2}^{\text{p}}$ and $T_{1/2}^{\text{b}}$ are the physical and biological half-lives of the radionuclide, respectively. From this equation, it can be seen that if physical half-life is much greater than biological half-life the effective half-life of a nuclide is determined primarily by its biological half-life. Similarly, if the biological half-life is much larger, the physical half-life dominates. The biological half-life of a nuclide should be sufficiently long to allow adequate nuclear activity being generated but also short to minimize radiation hazard.

**Differential Uptake**   The radionuclide should be taken by the normal or abnormal tissue of interest with high specificity. This will result in high contrast for best image quality.

**Toxicity**   An ideal radionuclide should be nontoxic.

## H.   Counting Statistics

Nuclear decay is a random process. The disintegration rate of a radioactive sample varies randomly, that is, the detected counts in successive measurements are not the same. However, the frequency distribution, defined as the probability density function of possible measurement outcome, follows the Poisson distribution given by

$$P(N, \lambda) = \frac{\lambda^N}{N!} \tag{3-12}$$

where $N$ is the possible outcome and $\lambda$ is the true value of the measurement, which can be estimated by the average value of the measurements

$$\lambda = \mathbf{N} = \frac{1}{n} \sum_{i=1}^{n} N_i \tag{3-13}$$

where $N_i$ is the $i$th measurement outcome and $n$ is the number of measurements. The variance of the Poisson distribution or of the measurement outcomes is given by

$$\sigma^2 = \lambda = \mathbf{N} \tag{3-14}$$

From Eqs. (3-12) and (3-14), we find that the relative standard deviation of the Poisson distribution is given by

$$\frac{\sigma}{\mathbf{N}} = \frac{1}{\sqrt{\mathbf{N}}} \tag{3-15}$$

Equation (3-15) indicates that the relative width of the Poisson distribution is inversely proportional to the square root of the average counts. In radionuclide imaging, $\sigma/\mathbf{N}$ can be related to the statistical noise fluctuations in the measured data and has important effects on the image quality.

Figure 134 shows the discrete Poisson distributions with $\mathbf{N} = 5$, 10, and 20. At low $\mathbf{N}$, the distribution is skewed. For $\mathbf{N}$ larger than about 15, the Poisson distribution is practically symmetric and has a shape almost identical to the Gaussian distribution with mean and standard deviation equal to $\mathbf{N}$ and $\sqrt{\mathbf{N}}$, respectively.

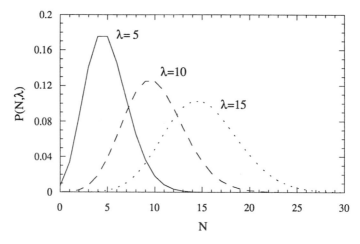

**Figure 134** Poisson's distributions with different true values.

# II.  Generation and Detection of Nuclear Emission

## A.  Nuclear Sources

Most naturally occurring radioactive isotopes are not useful in nuclear medicine because of their long half-lives and charged particle emissions. Due to these characteristics, artificially produced radioisotopes are exclusively used in radionuclide imaging. Artificially produced radioisotopes are created by bombarding stable isotopes with high-energy photons or nuclear particles. The stable isotopes undergo nuclear transformation and convert into unstable nuclei, or radioisotopes, which in turn transform into another stable isotope through radioactive decay.

Radioisotopes can be produced by nuclear reactors or by charged-particle accelerators. In nuclear reactors, neutrons from nuclear fission are captured by nuclei of the target material. New radioisotopes are generated through the neutron activation processes. The two common neutron activation processes are the $(n, \gamma)$ and the $(n, p)$ reactions. Examples of reactor-produced radioisotopes are $^{14}$C, $^{24}$Na and $^{32}$P. These radioisotopes tend to decay by emitting beta particles, and often have very low specific activity.

Alternatively, accelerators are used to accelerate charged particles such as protons and deuterons to very high energy. These high-energy charged particles can be directed to activate target material producing radioisotopes. There are a variety of charged-particles accelerators including Van de Graaff accelerators, linear accelerators, and cyclotrons. The cyclotron-produced radionuclides $^{11}$C, $^{13}$N, $^{15}$O, and $^{18}$F are positron emitters and are useful in PET imaging. They can be produced with high specific activity and can be labeled with a number of important physiological tracers. However, these radionuclides have short half-lives in the order of minutes. The need for an expensive on-site cyclotron to produce these radionuclides for clinical use has impeded the widespread use of PET imaging methods.

Finally, very high energy photons ($>10$ MeV) can be used to activate stable nuclei. Such a process is called photonuclear activation. Typical reactions are $(\gamma, n)$ and $(\gamma, p)$. This activation method tends to produce a low level of positron emitters that are not carrier-free, and it is seldom used in nuclear medicine.

## B.  Radionuclide Generators

A radionuclide generator is an apparatus that contains a parent–daughter radioisotope pair. The device allows separation and extraction of the daughter radioisotope from the parent. Figure 135 is a schematic diagram of a $^{99m}$Tc generator, sometimes referred to as a "radioactive cow." Inside the generator, the parent $^{99}$Mo is absorbed on an alumina ($Al_2O_3$) column or by alumina beads. The parent radioisotope $^{99}$Mo has a half-life of 2.5 days and decays according to the following reaction

$$^{99}\text{Mo} \longrightarrow {}^{99m}\text{Tc} + e^- + \bar{\nu}_e \qquad (3\text{-}16)$$

where $e^-$ and $\nu_e$ are electron and anti-electron neutrino, respectively. The daughter radioisotope is the metastable $^{99m}$Tc which has a half-life of 6 hours and decays

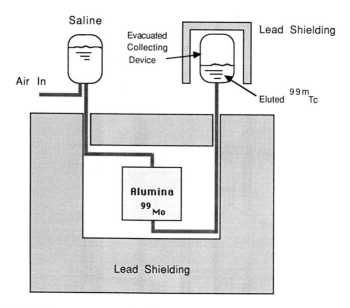

**Figure 135** Basic components of a $^{99m}$Tc generator.

to the stable nuclei $^{99}$Tc by emitting gamma radiation at 140 keV. Since $^{99m}$Tc does not bind with alumina, it can be eluted from alumina with saline. After elution, the $^{99m}$Tc activity is replenished in a few hours from the decay of $^{99}$Mo. Problems associated with technetium generators are related to "break through" of $^{99}$Mo and aluminium ions, which means that part of $^{99}$Mo and aluminium ions is also eluted with $^{99m}$Tc by saline. These "break through" problems should be avoided to minimize patient radiation dose and ensure purity of the daughter radioisotope.

## C. Nuclear Radiation Detectors

Generally, three types of detectors have been used to detect emissions from radioactive decays. They are ion collection detectors, scintillation detectors, and semiconductor detectors.

### 1. Ion Collection Detectors

A simplified schematic diagram of a typical ion collection detector is shown in Fig. 136. It consists of two electrodes (the negative and positive electrodes are called the cathode and anode, respectively) and a separated volume of air or gas of high atomic number such as xenon. An external voltage source is applied to the electrodes. The volume of air or gas acts as an insulator until it is ionized by radiations incident on the detector causing a short pulse of electric current in the detector circuit. A plot of the current $I_d$ as a function of the applied voltage is shown in Fig. 137.

When the applied voltage is set at zero, the current is zero even though ion pairs are being formed continuously as a result of the radiations. This is because the ion pairs do not have enough energy to separate and they recombine to form a neutral atom. Detectors operating in this range of applied voltage (or region 1)

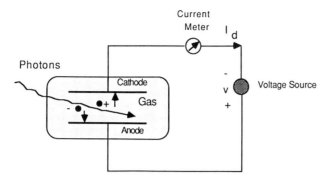

**Figure 136** Schematic diagram of an ion collection detector.

are known as ionization chambers. The ionization chambers are suited for detecting high-intensity radiations. The current produced is linearly proportional to the intensity of radiation at a certain voltage. In this respect, it is different from other counters whose response is proportional to the number of ionizing events.

As the applied voltage is increased, more and more electrons acquire enough energy to reach the anode, thereby resulting in an increase in current. Further increase of the applied voltage causes the electrons and the positive ions to gain enough energy to ionize other air or gas molecules. Therefore, a steeper increase in current is observed. Detectors operating in this region, or region 2, where the current is proportional to the incident radiation, are called proportional counters.

When the applied voltage is increased beyond region 2, a point will be reached where almost all of the molecules are ionized. An increase in applied

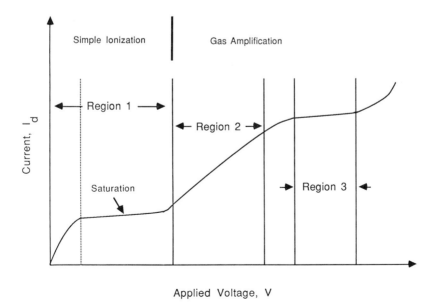

**Figure 137** Discharge current of an ion collection detector as a function of applied voltage.

voltage results in only a very small increase in current. Geiger–Muller (GM) counters operate in this region, or region 3. They are best suited for detecting a small amount of radioactivity because a single event will trigger an avalanche of ionization throughout the gas volume, independent of the type of nuclear radiation.

Further increase in the applied voltage beyond region 3 results in spontaneous ionization within the detector volume. The current will increase again in this so-called spontaneous discharge region. However, the spontaneous discharge region should be avoided because no useful information can be obtained.

Ion collection detectors have found limited use in nuclear medicine. The main reason is their poor detection efficiency for high energy γ rays. Also, the response time of these detectors is long compared to other types of radiation detectors.

## 2. Scintillation Detectors

Figure 138 shows the construction of a scintillation detector, which consists of a scintillation crystal such as NaI(Tl) for converting high-energy X- or gamma-ray photons into visible light photons. The NaI(Tl) is most popular in radionuclide imaging due to its high density of 3.69 gm/cm$^3$, relatively fast response time, ease in handling, and low cost.

Lead shielding is important to reduce detection of background radiation by the crystal. When an X- or gamma-ray photon is absorbed by the crystal, the intensity of light scintillations or the number of visible light photon emissions generated is proportional to the energy of the absorbed photon. These visible light photons are then converted into electrons by the photocathode. The photomultiplier tube (PMT) consists of a number dynodes at increasingly higher electric potential. The electrons from the photocathode are accelerated toward the anode through the stages of dynodes, each releasing an increasing number of secondary electrons. The total electron multiplication factor of a PMT is in the order of 6$^{10}$ for a 10-stage PMT. The magnitude of the final electric pulse from the PMT is proportional to the number of light photons incident on the PMT. This electric pulse is then amplified by the amplifier and analyzed by a pulse height analyzer (PHA) to determine the energy of the absorbed photon.

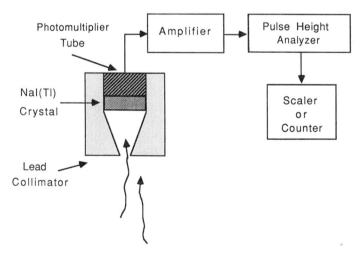

**Figure 138** Schematic diagram of a scintillation detector.

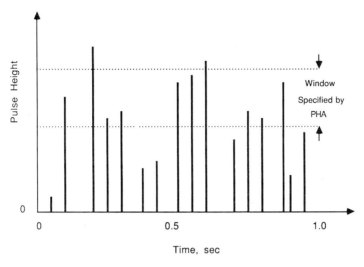

**Figure 139**  Output from pulse height analyzer (PHA).

Typical pulse heights detected by a PHA in a radiation detection experiment are shown in Fig. 139. By specifying a window, only electric pulses with certain heights for counting are selected. Since the pulse height is proportional to the energy of the absorbed photons, the pulse height window is also called the energy window. The ability of a scintillation detector system to select an energy window for photon detection is important in radionuclide imaging in rejecting scatter radiations.

It is sometimes necessary to determine the number of detected photons versus absorbed energy or pulse height. In this case, a multiple-channel analyzer (MCA) is often used. The MCA consists of digital electronics that sort the detected pulses into a number of channels (typically 256 or 512). Figure 140 shows the output from an MCA as an NaI(Tl) detector is exposed to a $^{99m}$Tc source. The count rates at the lower energy levels are due to Compton scattering of the gamma photons in the scintillation crystal. The spreading of the main energy peak at 140 keV is caused by the statistical variations in the numbers of the scintillation light photons generated in the crystal, the electrons released in the dynodes, and electrical noise in the PMT.

The spread or broadening of the energy peak is referred to as the energy resolution of the detector system. The energy resolution is usually defined as the full width at half-maximum (FWHM) of the measured energy peak, $\Delta E$. It is often expressed as a percentage of the photopeak energy $E_0$, that is,

$$\text{energy resolution} = \frac{\Delta E}{E_0} \times 100\% \qquad (3\text{-}17)$$

The energy resolution affects the ability of a radiation detector to distinguish gamma-ray photons with similar energies. It is useful in identifying radionuclides and in rejecting scatter radiation. The latter is important in radionuclide imaging to improve image contrast degraded by scatter radiation.

An important type of scintillation detector used in nuclear medicine is the well counter, shown in Fig. 141. It is designed to measure radiation from small radioactive samples. Due to their almost $4\pi$ geometry, well counters have high

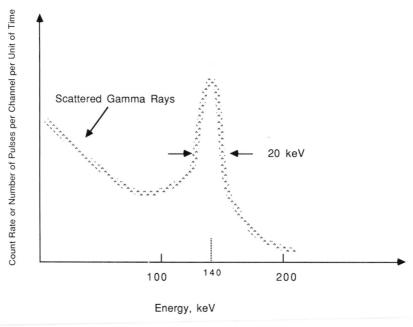

**Figure 140**  Output from multichannel analyzer due to $^{99m}$Tc source. Detector has energy resolution of 14%.

detection efficiency and are well suited for counting weak sources. For the detection of very weak radiation, liquid scintillation detectors can be used. In this type of detector, scintillation liquid is mixed with the radioactive sample. While in a solution, the sample can interact directly with the scintillation material for the highest possible detection efficiency.

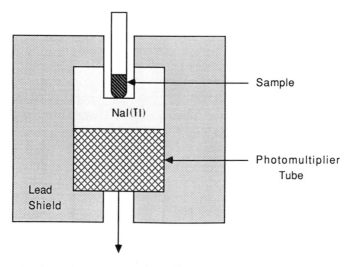

**Figure 141**  Physical construction of a well counter.

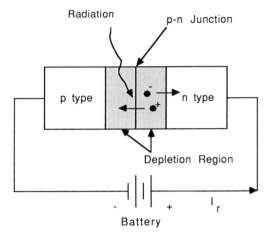

**Figure 142**   Schematic diagram of a solid-state detector.

## 3.   Solid-State Detectors

As shown in Fig. 142, when reversely biased, a depletion region characterized by the absence of free charge will be created near the p-n junction on a semiconductor such as silicon (Si) or germanium (Ge). A gamma-ray photon or energized particle can interact with the semiconductor material to form electron/hole pairs. With the application of the bias voltage, there will be an increase in the reverse current $I_r$ whose magnitude is directly proportional to the absorbed radiation energy.

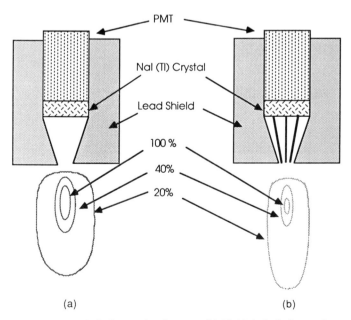

**Figure 143**   (a) Single-hole focused collimator. (b) Multiple-hole focused collimator.

Semiconductors have superior energy resolution in the order of a few keV and fast response time of about $10^{-9}$ sec. However, their applications in nuclear medicine are limited due to their small size and high cost. They are used primarily for charged-particle and gamma-ray spectrometers. Silicon detectors may be operated at room temperature but germanium can only be used at very low temperature ($-196°C$).

## D. Collimators

*SPATIAL RESOLUTION DEPENDS ALMOST SOLELY UPON COLLIMATORS.*

A collimator is used to limit the field of view of a radiation detector. Two common types of collimators are shown in Fig. 143. Figure 143 (a) shows a single-hole focused collimator for detecting gamma rays from a relatively large volume of radioactive source, whereas Fig. 143 (b) shows a multiple-hole focused collimator for detecting a more localized region of radioactivity. The dashed lines in the figures are isocount lines of the collimators indicating the shape of the field of view of the collimators. The focused collimator is useful in radionuclide imaging using rectilinear scanner systems. It is the major component that determines the spatial resolution and detection efficiency of the imaging system.

# III. Diagnostic Methods Using Radiation Detector Probes

In this section, applications of radiation detector probes in clinical diagnoses are discussed. These probes generally consist of a single scintillation detector system described in the previous sections without scanning mechanism. Valuable diagnostic information can be obtained by monitoring and evaluating the radioactivity from a organ of interest relative to a phantom or to a normal part of the organ or by following the activity as a function of time (Nudelman and Patton, 1980). A few examples are given in the following.

## A. Thyroid Function Test

The 24-hour uptake of radioactive iodine by the thyroid gland has been used for a number of years in the evaluation of thyroid function. A simple scintillation detector system as described in Fig. 138 can be used. A typical commercial thyroid uptake system is shown in Fig. 144. The thyroid requires iodine to produce hormones needed for regulating the metabolism of the body. A person with an underactive thyroid or hypothyroid has less iodine in the thyroid than normal. On the other hand, a person with an overactive thyroid or hyperthyroid has a higher concentration of iodine than normal. In the thyroid function test, approximately 300 kBq of $^{131}I$ contained in a liquid or a capsule is given to the patient. A sample with the same initial amount of $^{131}I$ is also prepared as reference. Twenty-four hours later the amount of $^{131}I$ is determined by counting the gamma particles emitted by the radioactive iodine for one minute. At the same time, the reference source is placed in a neck phantom and counted for the same period of time. The ratio of detected counts from the thyroid and from the phantom with the reference source gives the 24-hour uptake of iodine in the thyroid. For normal people,

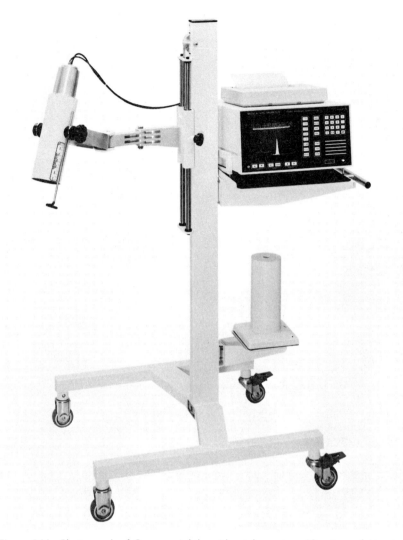

**Figure 144** Photograph of Commercial thyroid uptake system. (Courtesy of Atomic Products Corp., Shirley, New York).

the ratio is around 20%. If the ratio is greater than 60%, the patient is said to have a hyperthyroid. If it is less than 10%, the patient may have a hypothyroid.

## B. Renal Function Test

Renal function can be evaluated using radiation detector probes. For example, in a renal function test, seven MBq of $^{131}$I labeled hippuric acid is injected into the blood stream. The excretion of the radio-labeled hippuric acid by the kidneys is monitored by the radiation detector. The time–activity curve of the detected counts from the kidneys can be obtained by using a radiation detector positioned over the kidneys and is called a renogram. A typical example of a renogram is shown in Fig. 145. The usefulness of renal function analysis is illustrated in Fig. 146 for a patient with renal arterial stenosis in the left kidney.

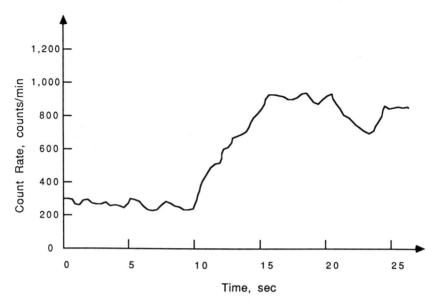

**Figure 145** Time-count rate curve for kidney or renogram following injection of [131]I labeled hippuric acid in patient blood stream as a function of time.

## C. Blood Volume Measurement

A technique using [131]I labeled albumin, based on the dye dilution technique, can be used to measure the blood volume accurately. About 200 kBq of the radioactive albumin solution is injected into a vein and after about 15 minutes a sample of blood is drawn and counted. If the sample is taken in a time interval much shorter compared to the half-life of the radionuclide, the blood volume can be determined using the following equation

$$V_b = \frac{A_0}{A_b} V_r \qquad (3\text{-}18)$$

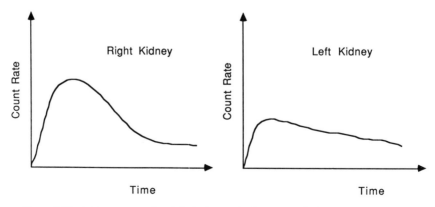

**Figure 146** (a) Renogram for right kidney, which is normal. (b) Renogram for left kidney with renal arterial stenosis.

where $V_b$ and $V_r$ are the volumes of blood and radionuclide containing liquid, and $A_0$ and $A_r$ are the activities of the radioactive liquid and of the withdrawn blood sample per unit volume, respectively.

# IV.  Radionuclide Imaging Systems

## A.  Rectilinear Scanner

Prior to the introduction of the scintillation or gamma camera, the rectilinear scanner was the principal imaging instrument used in radionuclide imaging. As shown in Figure 147, a typical rectilinear scanner consists of a scintillation detection system with a focused collimator, a mechanical scanning device, and a recorder system. The collimator acts like the lens in an optical imaging system. However, instead of focusing the gamma photons by refraction, the collimator is a passive device that allows gamma radiation to pass through the collimator hole properly to interact with the crystal. Other gamma photons are blocked by the lead shield or septa of the collimator. The collimator has major effects on the spatial resolution and detection efficiency of the rectilinear scanner. By scanning the collimated detector system over the radioactivity distribution in a rectilinear format, as shown in Figure 148, a two-dimensional detected image can be obtained. The detected events at each scanning position are recorded on a recording medium such as CRT or film to produce the recorded image for viewing and/or archiving. Depending on the recording medium, the recorded image consists of a pattern of dots or light intensity distribution. The density of dots or light intensity level corresponds to the relative concentration of radionuclide in the organ of interest. Figure 149 shows a typical image of the upper part of a human body obtained by a rectilinear scanner.

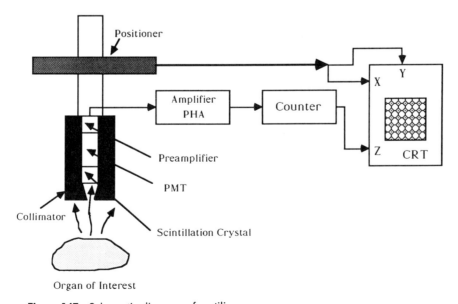

**Figure 147**  Schematic diagram of rectilinear scanner.

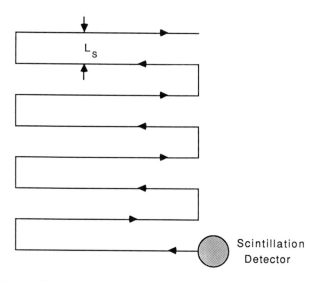

**Figure 148** Rectilinear motion of the scintillation detector of a rectilinear scanner.

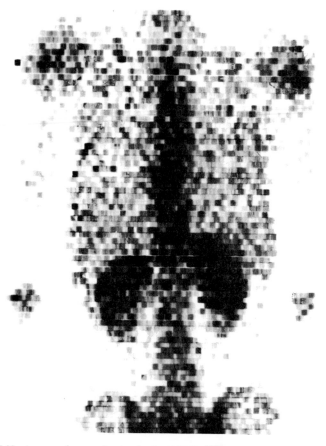

**Figure 149** Image of upper body obtained by rectilinear scanner.

The quality of the image obtained is related to the counting rate from the radioactivity distribution, and to the scanning speed and line spacing of the rectilinear scan. A parameter that can be used to describe image quality generated by a rectilinear scanner is the count information density ($D_{ci}$), which is defined as the number of counts recorded per unit scan area (counts/cm$^2$). It is related to the count rate (counts/min) $R_c$, the scan speed (cm/min) $S$, and line spacing (cm) $L_s$ by the following equation:

$$D_{ci} = \frac{R_c}{SL_s} \qquad (3\text{-}19)$$

From Eq. (3-19), we find that image quality can be improved by increasing $R_c$, or decreasing $L_s$ or $S$. However, the highest $D_{ci}$ attainable is limited by the trade-off between patient dose, which determines $R_c$, and total scan time, which influences the choice of $L_s$ and $S$. It has been found that $D_{ci}$ in the order of 800 counts/cm$^2$ is satisfactory for most situations.

The collimator has major effects on the spatial resolution and detection efficiency of a rectilinear scanner. The multihole focused collimator design provides optimal trade-off between spatial resolution and detection efficiency. As shown in Figure 150(a), by focusing all holes to a single point, the radius of view $R$ at a distance equal to the focal length is given by

$$R = \frac{2af}{d} \qquad (3\text{-}20)$$

where $a$ is the radius of a single collimator hole, and $d$ and $f$ are the thickness of the collimator and the focal length of the collimator holes. The radius of view provides a qualitative measure of the spatial resolution of a multihole focused collimator.

The spatial resolution of a multihole focused collimator can be expressed more precisely by the geometric transfer function given by Metz *et al.* (1974).

$$\mathrm{TF_g^{m.h.}}(\boldsymbol{\nu}) = \mathrm{TF_g^{s.h.}}(\boldsymbol{\nu}) \cdot H(\boldsymbol{\nu}, \theta, s) \qquad (3\text{-}21)$$

where $H(\boldsymbol{\nu}, \theta, s)$ is the two-dimensional Fourier transform of the hole array, $s$ is the spacing between the centers of adjacent collimator holes at the backplane, $\boldsymbol{\nu} = (\zeta, \xi)$ and $\theta$ is the orientation of the hole array with respect to the direction of the measured line spread function. Definitions of $s$ and $\theta$ are shown in Fig. 150(b). The geometric transfer function of a single hole $\mathrm{TF_g^{s.h.}}(\boldsymbol{\nu})$ is given by

$$\mathrm{TF_g^{s.h.}}(\boldsymbol{\nu}) = A[(z/d)\boldsymbol{\nu}] \cdot A[f(d+z)/d(d+f)\boldsymbol{\nu}] \qquad (3\text{-}22)$$

where $z$ is the source distance, and $A(\boldsymbol{\nu})$ is the Fourier transform of the collimator hole aperture function and is determined by the hole shape. The geometric transfer function provides a full description of the spatial resolution of the multihole focused collimator, which includes the configuration of the hole array and hole shape.

The geometric efficiency of a multihole focused collimator is given by

$$G = \frac{NA_1 A_2}{4\pi d^2} \qquad (3\text{-}23)$$

where $A_1$ and $A_2$ are the areas of the collimator hole at the front and back plane, and $N$ is the number of holes.

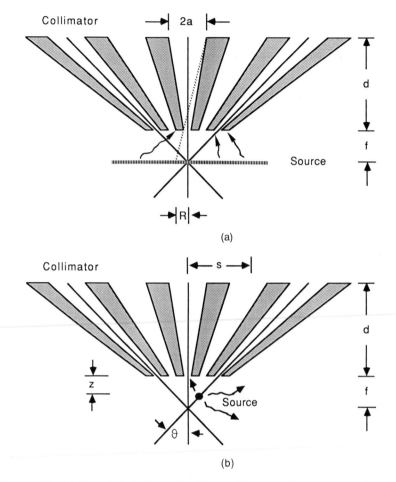

**Figure 150** (a) Multiple-hole focused collimator illuminated by a plane nuclear source. (b) Schematic diagram defining $s$ and $\theta$. (Dimensions of the collimator are not to scale.)

The improvement of spatial resolution can be achieved by decreasing the hole size or focal length or increasing the collimater thickness with a concurrent decrease of geometric efficiency. Hence, optimum collimator design involves trade-off between spatial resolution and detection efficiency. For detecting high-energy photons, single-hole focused collimators are sometimes used to minimize photon penetration through the collimator septa.

An extension of a rectilinear scanner with a single detector is the use of a linear array of scintillation crystals. Rectilinear scanner systems based on linear detector arrays provide better image quality in less scanning time. However, with the development of the large-area scintillation or gamma camera, rectilinear scanners are no longer used in modern nuclear-medicine clinics. Collimated detector arrays can be found in a few special SPECT system designs.

## B.  Scintillation Camera

The scintillation camera, also known as the gamma camera or the Anger camera, was first proposed and developed by Anger in the late 1950s. It is the most commonly used imaging device in radionuclide imaging today. By eliminating the rectilinear scanning motion, radioactivity distribution from a large area of the body can be imaged simultaneously with increased efficiency. It also allows functional imaging of radioactivity uptake and/or washout.

A typical commercially available scintillation camera is shown in Fig. 151. There are two major components of a scintillation camera, namely the camera head and the electronic processing unit. The camera head consists of the collimator and the scintillation crystal. Similar to the focused collimator for a rectilinear scanner, the camera collimator determines the spatial resolution and detection efficiency of the scintillation camera. Typical camera collimator consists of parallel holes that provide a one-to-one relation between the size of the object distribution and the image (Fig. 152). The characteristics of camera collimators will be discussed in more detail in a later section.

The scintillation crystal in a modern large field-of-view (LFOV) camera is 38 cm in diameter (in a jumbo camera it is in the order of 50 cm). The crystal thickness is a trade-off between the intrinsic resolution of the camera and detection efficiency of the incident photons. Most scintillation cameras are designed for imaging $^{99m}$Tc labeled pharmaceuticals with 140-keV low-energy $\gamma$ photon emissions, and a crystal thickness of 9.5 mm provides the best compromise.

Immediately behind the crystal is a hexagonal array of PMTs, as shown in Fig. 153. The number of PMTs affects the intrinsic resolution of the camera, and most LFOV cameras consist of 61 or more closely packed PMTs. The PMTs are coupled to the crystal directly or through light pipes. The output of each PMT is

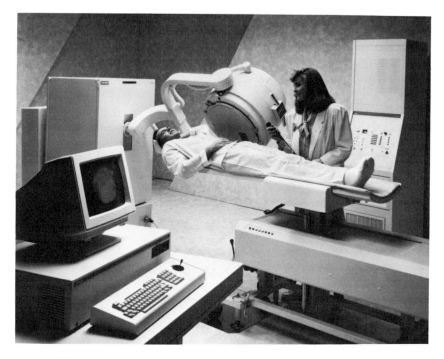

**Figure 151** Photograph of commercial gamma camera based single-photon emission computed tomography (SPECT) system. To produce a SPECT image, camera has to be translated. For non-SPECT imaging, camera position is fixed. (Courtesy of Siemens Gammasonics, Inc., Hoffman Estates, Illinois).

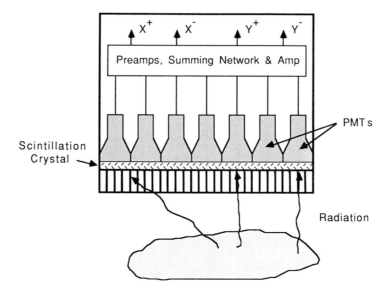

**Figure 152**   Gamma camera with parallel-hole collimator.

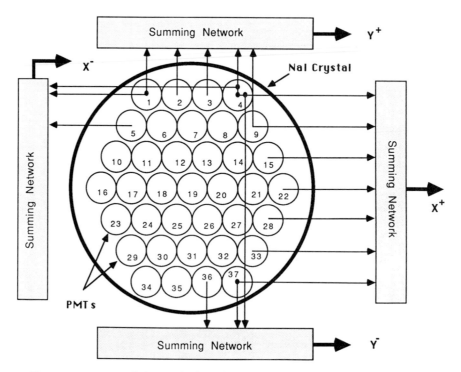

**Figure 153**   Matrix of photomultiplier tubes (PMTs) located behind the scintillation crystal in a gamma camera.

fed to a preamplifier for pulse amplification and shaping. The signals from the preamplifiers are combined into four composite signals $X^+$, $X^-$, $Y^+$, and $Y^-$ before further processing.

The block diagram of the electronic processing unit is shown in Fig. 154. The processing of a detected event can be understood by referring to Fig. 153. Suppose a scintillation event occurs below PMT #1. Most of the scintillation photons generated will be detected by PMT #1. Lesser amounts will be detected by PMT #2, #5, and #6. Still lesser will be detected by PMT #10, #11, #12, #7, and #3, and none will be detected by PMT #37. In the scheme shown, a pulse coming out of PMT #1 will contribute more to the sums appearing at $Y^+$ and $X^-$ and less to the sums appearing at $Y^-$ and $X^+$.

The four composite signals $X^+$, $X^-$, $Y^+$, and $Y^-$ are amplified by the summing amplifiers whose output amplitudes are subsequently adjusted by the attenuators. Beyond this point, each of the four composite signals is split into two separate paths. In one path, the four composite signals are summed. The amplitude of the summed pulse is proportional to the intensity of the scintillation or the energy of the absorbed photon. The summed pulse is fed into the pulse height analyzer with a preset energy window. If the amplitude of the summed pulse is smaller or larger than the preset energy window, no further processing will occur. If the amplitude falls within the energy window, the summed pulse will turn on a gate that triggers the line amplifiers and wave-shaping circuits. The wave-shaping circuits

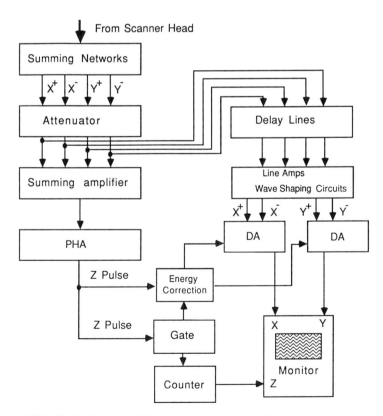

**Figure 154**  Block diagram of electronic processing unit in a gamma camera.

lengthens and shapes the narrow composite pulses. The processed pulses are then applied to the differential amplifiers (DA), which perform the following function:

$$\Delta x = -\frac{g}{Z}(X^+ - X^-) \tag{3-23}$$

and

$$\Delta y = -\frac{g}{Z}(Y^+ - Y^-) \tag{3-24}$$

where $g$ is the amplifier gain and $Z$ represents the pulse height. The DAs remove the energy dependence of the position signals. The signals $\Delta x$ and $\Delta y$ are used as $x$ and $y$ position signals for the display monitor to indicate location of the absorbed photon.

Figure 155 is an image of the kidneys obtained by a scintillation camera, encircled by bright lines (see color plate 3).

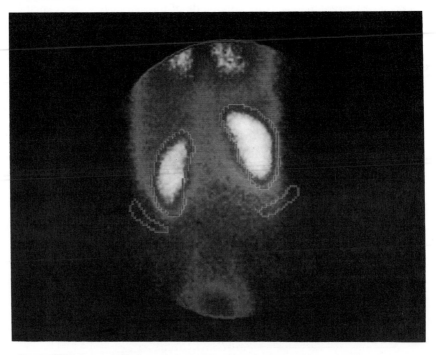

**Figure 155** Image of kidneys, which are outlined, obtained by a gamma camera (courtesy of Siemens Gammasonics, Inc., Hoffman Estates, ILL).

The intrinsic resolution of a scintillation camera is defined as the spatial resolution of the system without the collimator. A qualitative measurement of the intrinsic resolution can be obtained by using a bar phantom, shown in Fig. 156. The bar phantom consists of a square sheet of plastic with four sets of lead bars imbedded in it. The bars in each set have the same width and are separated by gaps equal to the width. The bar phantom is placed right in front of the crystal

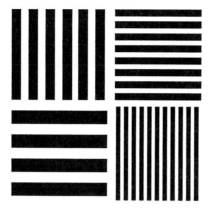

**Figure 156** Bar phantom.

and irradiated by a uniform flat radioactive source or a point source place at a large distance away from the camera. The radioactive source is usually made of $^{57}$Co or $^{99m}$Tc. An image of the bar phantom obtained by a scintillation camera is shown in Fig. 157.

Another important characteristic of the scintillation camera is the field uniformity. Ideally the response of a camera to a uniform irradiation should be uniform across the camera face. However, in practice, this is seldom the case and image uniformity can vary by as much as ±10% over the entire crystal surface. In other words, the number of counts displayed per unit area may vary by as much as 20% from one point to another when the camera face is uniformly irradiated. As recommended by NEMA (National Electrical Manufacturers Association) (NEMA, 1980), the intrinsic uniformity of a scintillation camera

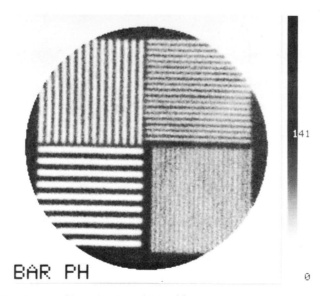

**Figure 157** Image of bar phantom obtained by a gamma camera.

should be expressed as the "integral" and "differential" uniformity. Integral uniformity is defined as

$$\text{Integral uniformity} = \pm 100 \left( \frac{\text{max} - \text{min}}{\text{max} + \text{min}} \right) \qquad (3\text{-}25)$$

where max and min are the maximum and minimum pixel counts in an image obtained from a uniform irradiation. Better uniformity corresponds to smaller integral uniformity.

The differential uniformity determined from a sliding group of 5 pixels is given by

$$\text{Differential uniformity} = \pm 100 \left[ \frac{\text{max. diff. (high} - \text{low)}}{\text{high} + \text{low}} \right] \qquad (3\text{-}26)$$

where max. diff. is the maximum difference between the high and low values of the 5 pixel sliding group over the image. It represents the worst-case rate of change across the uniformity image.

The stationary scintillation camera provides a major advance in radionuclide imaging instrumentation over the rectilinear scanner. The modern scintillation camera has intrinsic resolution in the order of 4 mm. With advanced electronic processing units, integral and differential uniformity in the order of ±5% can be achieved. By interfacing the scintillation camera to a digital computer, functional images can be acquired in specified time sequences. The dynamic data acquisition methods are especially useful in cardiac and renal imaging, which have become the major applications of radionuclide imaging techniques.

Similar to the rectilinear scanner, the collimator plays an important role in the performance of a scintillation camera. The parallel-hole collimator (Anger, 1964) is by far the most common type of collimator used in radionuclide imaging. As shown in Fig. 158(a), it consists of an array of holes separated by parallel lead septa. The collimator holes are oriented perpendicular to the surface of the crystal. As a result, the size of the image and of the radionuclide distribution have a one-to-one radio. The collimator has major effects on the spatial resolution and detection efficiency of a scintillation camera.

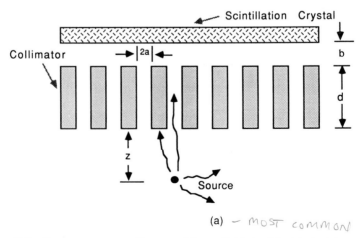

(a)  — MOST COMMON

**Figure 158** Gamma camera collimators: (a) parallel hole, (b) pinhole, (c) converging hole, and (d) diverging hole. (Dimensions of the collimator are not to scale.)

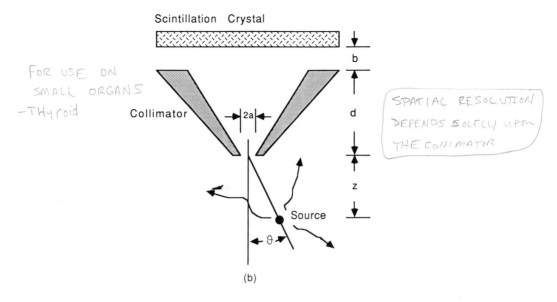

FOR USE ON
SMALL ORGANS
-THYROID

SPATIAL RESOLUTION
DEPENDS SOLELY UPON
THE COLLIMATOR

(b)

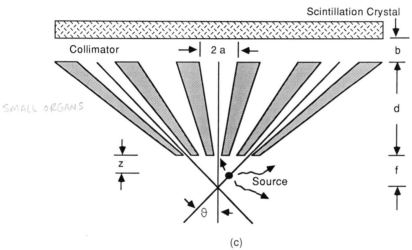

SMALL ORGANS

(c)

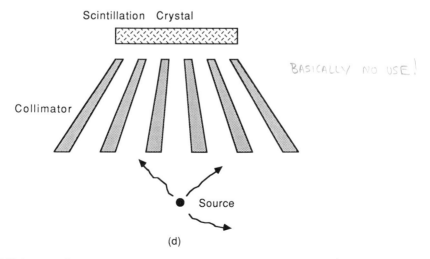

BASICALLY NO USE!

(d)

**Figure 158** *(continued)*

The spatial resolution of a parallel-hole collimator is related to the width of the average intensity distribution obtained from a point source placed at a distance $z$ from the collimator face (full width half maximum, or FWHM), that is,

MOST COMMON

$$R = \frac{2a(d + z + b)}{2(d + z)} \quad \text{PARALLEL HOLE COLL.} \tag{3-27}$$

where $2a$ and $d$ are the diameter of collimator holes and thickness of the collimator, and $b$ is the gap between the back of the collimator and the image plane inside the crystal. A more precise description of the spatial resolution of a parallel-hole collimator is given by the geometric transfer function (Metz et al., 1980)

$$TF_g(v) \simeq \left| A\left[ \frac{(d + z + b)}{d} v \right] \right|^2 \tag{3-28}$$

where $A(v)$ represents the normalized Fourier transform of the aperture function of a single hole and is determined by the shape of the collimator hole.

The geometic efficiency of a parallel-hole collimator is given by

YOU WANT G BIG

$$G = G_0\left( \frac{A_{\text{open}}}{A_{\text{unit}}} \right) \tag{3-29}$$

where

$$G_0 = \frac{A_{\text{open}}}{4\pi d^2} \tag{3-30}$$

is the geometric factor, $A_{\text{open}}$ is the open area of the hole aperture at the backplane, and $A_{\text{unit}}$ is the area of a unit cell of the hole array that includes a hole aperture and the septal material around it. Equations (3-29) and (3-30) show that the geometric efficiency of a parallel-hole collimator is independent of source distance. This can be understood by realizing that the decrease in detection efficiency due to the inverse square law is compensated exactly by an increase of crystal area that is seen through the collimator holes. From Eqs. (3-27), (3-29), and (3-30), we find that the geometric efficiency $G$ is approximately proportional to the square of the spatial resolution $R$. This implies that a fourfold increase in geometric efficiency will be accompanied by a twofold degradation in spatial resolution. The trade-off between geometric efficiency and spatial resolution of the camera collimator is an important consideration in radionuclide imaging.

Other types of collimators have been used with the scintillation camera. They include the pinhole, converging-hole, and diverging-hole collimators as shown in Fig. 158. These collimator types are designed for different clinical applications and are characterized by various trade-offs between spatial resolution, geometric efficiency, and field of view (Tsui, 1988).

The pinhole collimator shown in Fig. 158 (b) is used primarily for high-resolution imaging of small organs (e.g., thyroid) at close distances. The collimator provides a magnified image of the object, the magnification factor depending on the ratio $z/d$, where $z$ is the source distance and $d$ is the collimator thickness. The spatial resolution of a pinhole collimator can be given by

$$R = \frac{2a(z + d + b)}{d} \tag{3-31}$$

where $2a$ is the size of the pinhole. The geometric efficiency of a pinhole collimator is given by

$$G = \frac{4a^2 \cos^3 \theta}{16z^2} \tag{3-32}$$

where $\theta$ is the angle between the axis of the collimator and the line joining the center of the pinhole and the point source. Typically the diameter of the pinhole is in the order of a millimeter and $d$ is between 20 and 25 cm.

The converging-hole collimator shown in Fig. 158 (c) is used to image small organs. The magnification factor of an object between the focal point and the face of the collimator is $(f + d + b)/(f - z)$ where $z$ is the source distance, $d$ and $f$ are the thickness and focal length of the converging-hole collimator, respectively, and $b$ is the gap between the back of the collimator and the image plane inside the crystal. The spatial resolution of a converging-hole collimator can be described by

$$R = \frac{2a(d + z + b)}{d} \cdot \frac{1}{\cos \theta} \cdot \left(1 - \frac{d/2 + z}{f - z}\right) \tag{3-33}$$

CONVERGING HOLES

where $\theta$ is the angle between the axis of the collimator and a given collimator hole.

The geometric efficiency of a converging-hole collimator is given by

$$G = \left(\frac{A_{open}}{4\pi d^2}\right) \cdot \left(\frac{A_{open}}{A_{unit}}\right) \cdot \frac{f}{f - z} \qquad 0 \le z < f \tag{3-34}$$

where $A_{open}$ is the open area of the hole aperture at the backplane and $A_{unit}$ is the area of a unit cell of the hole array that includes a hole aperture and the septal material around it.

Diverging-hole collimators, shown in Fig. 158(d) are used with a small scintillation camera to image large areas of interests. They are seldom found in modern nuclear medicine clinics, because of the popularity of LFOV cameras.

# V. New Radionuclide Imaging Methods

Conventional radionuclide imaging using projection imaging techniques suffers from low image contrast due to overlapping structures. The advances in emission computed tomography (ECT) techniques have brought radionuclide imaging into a new era. In the following, the development of three important new radionuclide imaging methods, namely longitudinal section tomography, single-photon computed tomography, and positron computed tomography will be presented.

## A. Longitudinal Section Tomography

Longitudinal-section tomography techniques have been developed for radionuclide imaging. A rectilinear scanner with highly focused collimator is the simplest form of longitudinal-section tomography scanner. The scanner image contains a fo-

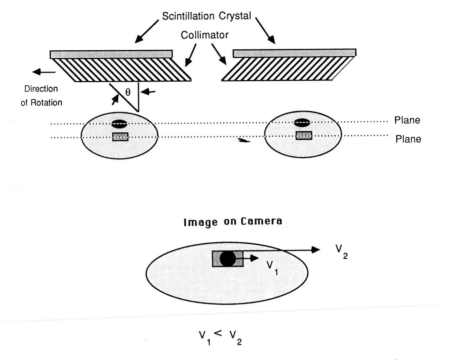

**Figure 159**    Longitudinal-section tomography using a scintillation camera with a rotatable slant-hole collimator.

cused image of the structures in the focal plane of the collimator and a blurred image of structures that are outside the focal plane. The image of the focal plane can be further enhanced by using multiple detectors or flexible coverging detectors that contain a scintillation solution (Cassen, 1965; Nudelman and Patton, 1980).

Longitudinal-section tomography can be achieved using a scintillation camera. One technique uses a rotatable slant-hole collimator as shown in Fig. 159 (Muehllehner, 1971). The projection of an object structure onto the image plane depends on the angle of the slant-hole with respect to the collimator axis, $\theta$. As the angle $\theta$ is changed, the position of the projection will move correspondingly. The speed of this movement is related to the distance between the object plane and the camera. The larger the distance, the greater the speed. If an image recording device is moved in synchrony with the motion of the projected image of object structures on the plane of interest, the recorded image will contain a focused image of the structures in the plane of interest and blurred images of structures in other planes.

A scintillation camera with a focused collimator can also be used to obtain tomograms of longitudinal sections. A multiplane tomographic scanner was developed by Anger (Anger, 1971) and marketed by Searle as the Pho/Con scanner system. The unit consists of two 23.6-cm diameter scintillation cameras, each fitted with a highly converging hole collimator. A simplified schematic diagram of the tomography system is shown in Fig. 160. The principle of tomography imaging used in this device is similar to conventional X-ray tomography. By scanning the dual scintillation cameras, the images of object structures in differ-

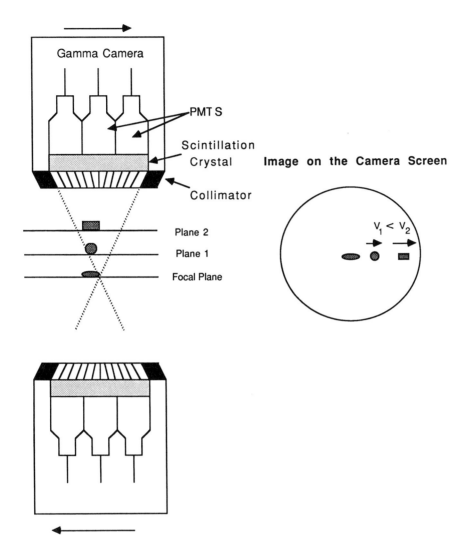

**Figure 160**  Longitudinal-section tomography using a scintillation camera with a focused multiple-hole collimator.

ent planes in the field of view of the collimators move across the crystal surface. The velocity of the image of an object from a given plane is proportional to the distance between the object and the focal plane. In Fig. 160 the image of the rectangular structure obtained by the top camera moves faster than the circular structure. By adjusting the relative speed of the recording device with respect to the motion of the cameras, the Pho/Con scanner is capable of recording 12 selected longitudinal-section tomograms.

As in conventional X-ray tomography, longitudinal-section tomography is inherently a limited-angle reconstruction problem. As in any other limited-angle reconstruction situations, a longitudinal-section tomogram consists of a focused image from a selected plane and blurred images of structures from other planes. Although the contributions from planes other than the focal plane are distorted or blurred, they may still be significant and lead to overall degradation in image quality, especially in lowering image contrast. The limitations of longitudinal-section tomography is largely eliminated in transverse-section tomography where projection images from 360° views around the object are used in image reconstruction. The reconstruction techniques are similar to those used in X-ray CT.

## B.  Single-Photon Emission Computed Tomography (SPECT)

The basic principles of transverse-section tomography or single-photon emission computed tomography (SPECT) are very similar to X-ray CT. By dividing a transverse section into a matrix of small compartments or voxels. The radioactivity of each voxel can be computed from projection data obtained from full 360° around the patient (Fig. 161). The projection data can be obtained by translating and rotating arrays of multiple detectors (multidetector approach) or by rotating one or more scintillation cameras around the patient (camera-based approach).

The SPECT method was first investigated by Kuhl and Edwards (Kuhl and Edwards, 1963), years before the development of X-ray CT. The original SPECT system, the Mark IV scanner, consisted of 4 linear detector arrays arranged to form a square opening as shown in Fig. 162 (Kuhl, 1976). Each detector array consisted of eight NaI(Tl) detectors. By offsetting the detector arrays by one-fourth of the detector spacing, a total of 32 linear samples can be obtained by rotating the detector array through 360°. The multidetector SPECT system approach was further modified and developed by a number of investigators (Patton *et al.*, 1973; Kuhl *et al.*, 1976). Commercial systems include those marketed by Medimatic Corporation (Stokeley *et al.*, 1980). Other multidetector SPECT systems use detectors arranged in a ring geometry. Examples are the SPRINT system developed at the University of Michigan (Williams *et al.*, 1979) and a commercial system by Shimadzu Corporation (Kanno *et al.*, 1981). Although multidetector

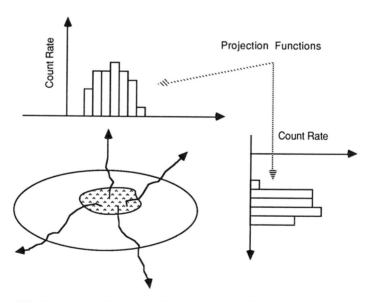

**Figure 161**   Projections of radioactivity in a body at different angles.

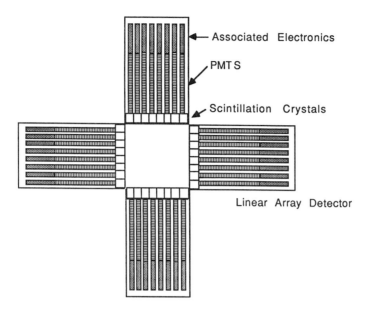

**Figure 162** Schematic diagram of Mark IV SPECT scanner developed by Kuhl (Kuhl, 1976).

SPECT systems have high count-rate capability, the systems require many components and are expensive to construct and maintain. Also, most of these systems are capable of imaging a single slice.

The camera-based SPECT system was first studied by Budinger (Budinger and Gullberg, 1974) with a stationary camera; the projection views were obtained by rotating the patient. The modern SPECT system using a rotating camera was developed independently by Keyes (Keyes *et al.*, 1977) and Jaszczak (Jaszczak *et al.*, 1977). Today all major medical imaging system manufacturers offer SPECT systems based on a single rotating camera. A single-camera SPECT is shown in Fig. 151. SPECT systems consisting of two, three, and four scintillation cameras are commercially available.

The reconstruction algorithms discussed in Section III-G-2 of Chapter 1 can all be used in SPECT image reconstruction. There are, however, differences between the X-ray CT and SPECT techniques. First of all, an X-ray CT image depicts the attenuation coefficients whereas a SPECT reconstructed image represents radioactivity distribution inside the patient. Furthermore, in SPECT the photon emissions are attenuated by the intervening tissues before reaching the detector and forming the projection data. Hence, to reconstruct a quantitatively accurate estimate of the radioactivity distribution, the attenuation effects must be compensated for.

In general, if the attenuation distribution is unknown, the reconstruction problem involves solving for two sets of unknown quantities, namely, the radioactivity and attenuation coefficient in each voxel from the projection data. Analytic solution to the above reconstruction problem has not been found. Special reconstruction methods using iterative algorithms have been applied in cinical situations where the nonuniform attenuation distribution can be determined from the transmission CT method (Malko *et al.*, 1986; Tsui *et al.*, 1989).

In situations where the attenuation coefficient can be assumed to be constant throughout the section to be reconstructed, a number of attenuation compensation methods have been developed. These compensation methods can be categorized into the intrinsic compensation techniques (Tretiak *et al.*, 1980; Gullberg and Budinger, 1981), algorithms that preprocess the projection data (Budinger *et al.*, 1979), and algorithms that postprocess the reconstructed image (Chang *et al.*, 1978).

In the preprocessing method, the attenuation effect is compensated for by taking the geometric or arithmetic mean of pairs of projection data that are 180° apart. For example, the basis for using the geometric mean can be explained by considering Fig. 163 where a radioactive source is located along the line joining the two detectors I and II. Suppose that the thickness of the radioactive source and of the body are $t$ and $T$, respectively. The source is at a depth $d$ from the surface closest to detector I. Then the detection counts in detectors I and II are given by (Sorenson, 1974)

$$C_1 = S_1 \cdot \sigma \cdot \exp(-\beta d) \cdot F(\beta, t) \qquad (3\text{-}35)$$

and

$$C_2 = S_2 \cdot \sigma \cdot \exp[-\beta(T - d)] \cdot F(\beta, t) \qquad (3\text{-}36)$$

respectively, where

$$F(\beta, t) = \frac{\sinh\left(\dfrac{\beta t}{2}\right)}{\left(\dfrac{\beta t}{2}\right)} \qquad (3\text{-}37)$$

is a slowly increasing function of the source thickness. For source thickness less than half the body thickness, $F(\beta, t)$ can be considered equal to 1 (Sorenson, 1974).

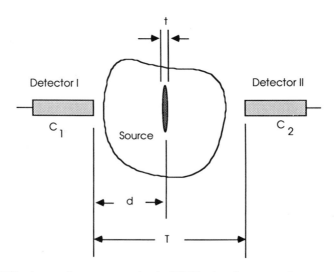

**Figure 163** Attenuation compensation in SPECT using the geometric mean.

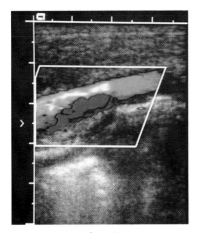

Plate 1

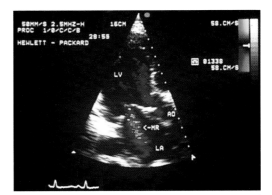

Plate 2

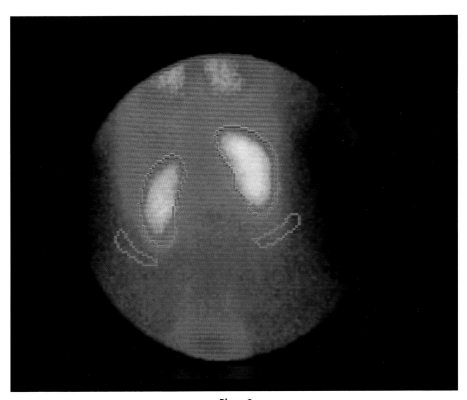

Plate 3

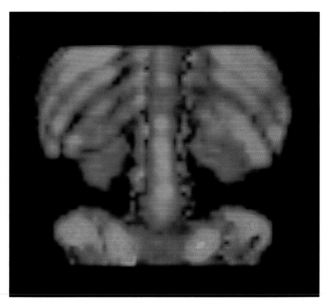

Plate 4

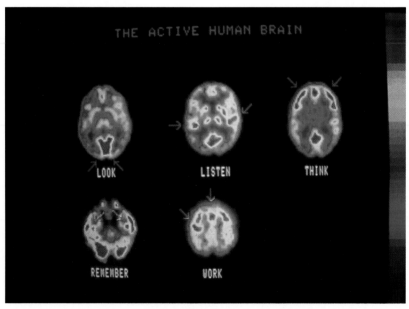

Plate 5

In Eqs. (3-35), (3-36) and (3-37), $\beta$ is attenuation coefficient in $\text{cm}^{-1}$, $S_1$ and $S_2$ are the sensitivity of the two detectors in counts-$\text{cm}^2$/sec-mCi, $\sigma$ is the radioactivity

$$(C_1 C_2)^{1/2} = \sigma (S_1 S_2)^{1/2} \exp(-\beta T) F(\beta, t) \tag{3-38}$$

The importance of Eq. (3-38) is that it shows the geometric mean of the detected counts $C_1$ and $C_2$ is independent of the source depth $d$, a quantity that is difficult to estimate from the projection data. For a point source where $t = 0$, the factor $F(\beta, t)$ equals 1 and the attenuation effects can be compensated for by the geometric mean as long as the body thickness $T$ can be estimated. For $t \neq 0$, the factor $F(\beta, t)$ is a slowly increasing function of $t$.

The geometric mean method is accurate only for a single source of radioactivity. When multiple sources are present, the analysis is more complicated and the Eq. (3-38) must be modified for the multiple source distribution (Tsui *et al.*, 1981; Sorenson and Phelps, 1987).

Most X-ray CT systems are single-slice units, that is, data are acquired and reconstructed from one transverse section at a time. In SPECT imaging using camera-based systems, the acquired data from each projection view are two-dimensional. Reconstruction of the two-dimensional projection data produces multiple transverse images. Hence, SPECT is a three-dimensional imaging technique. The volume of three-dimensional data obtained from the reconstruction can be displayed in transverse, sagittal, or coronal sections. Also, oblique sections with any orientation can also be reconstructed from the volume of data. Figure 164

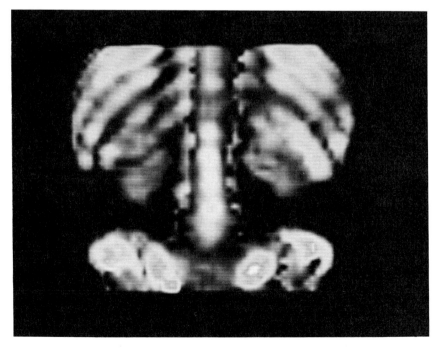

**Figure 164** Coronal section of mid-portion of lumbar spine following intravenous injection of $^{99m}$Tc labeled phosphate obtained by camera-based SPECT. (Courtesy of Siemens Gammasonics, Inc., Hoffman Estates, Illinois).

(see color plate 4) shows a SPECT coronal section of the mid-portion of lumbar spine following intravenous injection of $^{99m}$Tc labeled phosphate.

The counting statistics of SPECT image data are limited by the amount of radioactivity that can be safely administered into the patient. As we have discussed earlier, an increase of the detection efficiency of a collimated camera system has to be traded for poorer spatial resolution. Similar to planar imaging, the trade-off between detection efficiency and spatial resolution in SPECT imaging results in spatial resolution of about 10 to 15 mm for SPECT systems using a single rotating camera. The trade-off can be improved by using a specially designed collimator such as the fan beam (Jaszczak *et al.*, 1979; Tsui *et al.*, 1986) or cone beam collimators (Jaszczak *et al.*, 1988). However, the clinical use of a cone-beam collimator in SPECT imaging requires further research in the reconstruction to avoid image artifacts and distortions. Furthermore, new commercial SPECT systems are now available with multiple cameras. A commercial example is Picker's 3-head Prism SPECT imager. The increase in detection efficiency offered by these systems can be traded for improved spatial resolution. By combining the use of fan-beam collimators with a multicamera SPECT system, reconstructed images of the brain with spatial resolution in the order of 7 mm can be achieved.

Similar to the conventional nuclear medicine imaging, the spatial resolution of SPECT images depends largely on the collimator design used in the detector. Within a SPECT reconstructed image, the spatial resolution is about the same as the spatial resolution of the detector system at a distance equal to the radius of rotation. The main advantage of SPECT over conventional nuclear medicine imaging is that image contrast is substantially improved without the superposition of overlying and underlying distribution of radioactivity. The increase in image contrast is the main reason for the acceptance of SPECT in clinical nuclear medicine practice.

The clinical applications of SPECT have been mainly in the detection of tumor, in assessing myocardial infarction, and in assessing blood perfusion in the brain. The typical imaging time using a single-camera SPECT system is in the order of 30 minutes. As discussed earlier, tremendous progress has been made in improving the trade-off between detection efficiency and spatial resolution in SPECT imaging techniques.

These advances are accompanied by progress in the development of new radio-labeled pharmaceuticals for SPECT imaging. The concurrent development of new radiopharmaceuticals, instrumentation, and image-processing techniques has brought SPECT from research laboratories into clinical practice.

## C. Positron Emission Tomography (PET)

Although the potential of using positron emitters in medical imaging was suggested as early as 1951 (Wrenn *et al.*, 1951), they were not used until 1962 (Rankowitz, 1962). The excitement about positron tomography (PET) was created by the fact that most of the elements found in the human body have positron-emitting radioisotopes. Examples are $^{11}$C ($T_{1/2} = 20.5$ min), $^{13}$N ($T_{1/2} = 10$ min), $^{15}$O ($T_{1/2} = 122$ sec) and $^{18}$F ($T_{1/2} = 110$ min). By labeling active body constituents with these positron emitters, we are able to study *in vivo* physiological and metabolic functions, which is impossible using other techniques.

Furthermore, soon after being emitted a positron interacts with materials and combines with (or annihilates) an electron, typically within 1 to 3 mm range from the emission site, to form two gamma photons with 511-keV energy traveling in opposite directions. Two scintillation detectors with a fast coincidence circuit can be used to detect the two photons, as shown in Fig. 165. The function of the coincidence circuit is to allow a detected signal to be registered only if two events are detected within a certain time interval. It also acts as an "electronic collimator" that specifies the region of view of the coincidence detector systems. The detection efficiency of the electronic collimator is much higher than that of the passive collimators used in SPECT. By using a large number of coincidence detectors, PET provides a unique technique for radionuclide imaging that is inherently more sensitive than SPECT. Another advantage of PET is that the attenuation effects are much smaller than SPECT due to the higher-energy photons and are independent of the origin of the emission.

Coincidence detection is affected by the coincidence resolving time of the detectors. The detection of two photons from two different positron annihilations within the coincidence resolving time results in random coincidences. The random coincidence rate is given by

$$R_\tau = 2\tau \cdot S_1 \cdot S_2 \tag{3-39}$$

where $S_1$ and $S_2$ are the single channel noncoincidence count rates in the two detectors, and $\tau$ is the coincidence resolving time. Suppose $S_1$ and $S_2$ are the same and equal to $S$, the ratio of random coincidence rate $R_\tau$ to true coincidence rate $R_t$ is given by

$$\frac{R_\tau}{R_t} = \frac{2\tau S}{k_f} \tag{3-40}$$

where $k_f = \dfrac{R_t}{S}$ is the fraction of true coincidence over single rates and is usually a small number. From Eq. (3-40), we see that the random coincidence rate can be reduced by a shorter resolving time or detection at lower single rates.

The designs of PET systems were pioneered by a number of investigators (Robertson *et al.*, 1973; Ter-Pogossian *et al.*, 1975; Cho *et al.*, 1976; Thompson *et al.*, 1978; Brownell *et al.*, 1979; Derenzo *et al.*, 1979). As an example, Fig. 166 shows a schematic diagram for the ECAT II PET system (Williams *et al.*,

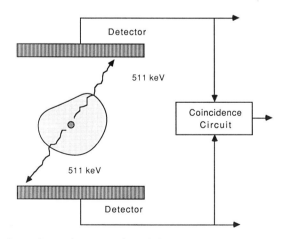

**Figure 165**  Coincidence detection of annihilation radiation.

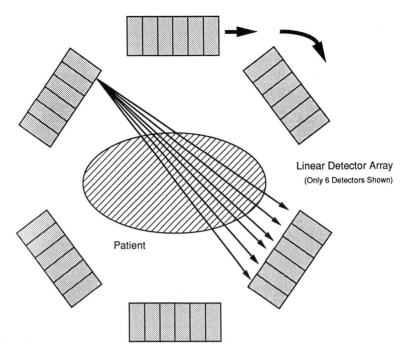

**Figure 166** Schematic diagram of ECAT II PET system (Williams *et al.,* 1979).

1979). It consists of 66 NaI(Tl) detectors arranged in a hexagonal array. Each bank consists of 11 detectors. The system produces 363 coincidence lines. The complete gantry linearly translates over a distance of 3.5 cm and then rotates in 5° incremental angle up to 60°. An updated version of this system is commercially available.

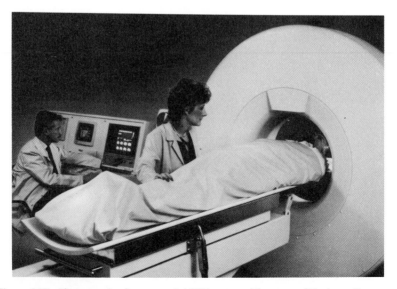

**Figure 167** Photograph of commercial PET system. (Courtesy of Positron Corp., Houston, Texas).

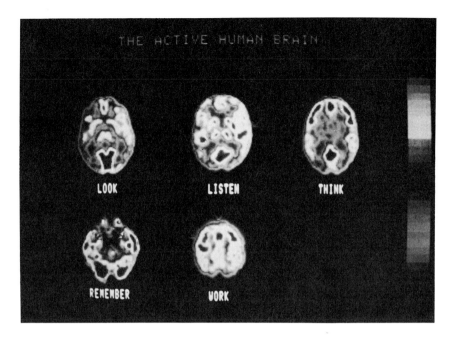

THE ACTIVE HUMAN BRAIN

LOOK          LISTEN          THINK

REMEMBER          WORK

**Figure 168**  PET images of an active brain under various states following intravenous injection of $^{18}$F labeled deoxy-glucose. (Courtesy of CTI PET Systems, Inc., Knoxville, Tennessee).

Several important developments have further improved the performance of PET imaging technique. Examples are the development of multislice PET system (Ter-Pogossian *et al.*, 1978), the use of BGO scintillator (Cho and Farukhi, 1977), and the development of time-of-flight PET (TOF-PET) (Ter-Pogossian *et al.*, 1981) using fast scintillators such as godolium orthosilicate (GSO), cesium flouride ($C_sF$), and barium flouride ($BaF_2$) (Laval *et al.*, 1983), and a block crystal design to reduce the number of electronic components (Casey and Nutt, 1986; Eriksson *et al.*, 1987). A number of PET systems with different designs are commercially available today. One of such systems is shown in Fig. 167. PET images of the brain are shown in Fig. 168 (see color plate 5) following intravenous injection of $^{18}$F labeled deoxy-glucose.

The major disadvantage of PET imaging is that all positron emitters have short half-lives. This means that the radionuclides have to be produced on site. Cyclotron units capable of producing positron emitters are now commercially available for installation in hospital sites. However, the additional costs of a cyclotron unit is a major deterrent to the widespread use of PET.

# VI.   Characteristics of Radionuclide Images

Nuclear image quality can be assessed by the same measures discussed in Section V of Chapter 1 for X-ray, namely spatial resolution and image contrast and noise.

## A.  Spatial Resolution

The spatial resolution of a radionuclide imaging device is largely determined by the collimator. Other factors include photon energy, system uniformity, characteristics of the PMTs, and patient motion. A qualitative measure of system resolution can be obtained from an image of the bar phantom shown in Fig. 156. Quantitative measures of spatial resolution include the point-spread function, line-spread functions, edge-spread function, and optical transfer function and modulation transfer function (Metz and Doi, 1979). These descriptors have been applied in medical imaging system analysis using linear system theory.

## B.  Image Contrast

Image contrast is affected by the distribution of the radionuclide used, scatter radiation, and penetration through the collimator septa. Contrast resolution of an imaging system can be measured by the contrast efficiency function.

Suppose the profile of radioactivity distribution from a phantom filled with a radioactive solution is give in Fig. 169. The normalized object contrast is given by

$$C_0 = \frac{A_l - A_b}{A_l} \tag{3-41}$$

where $A_l$ and $A_b$ are the radioactivity levels at the lesion and at the background, respectively. The corresponding normalized image contrast, $C_i$, obtained at the output of the imaging device, is given by

$$C_i = \frac{D_l - D_b}{D_l} \tag{3-42}$$

where $D_l$ and $D_b$ are the lesion and background optical density if the image is recorded on film, and are respective gray levels if a digital displaying device is used. The contrast efficiency is defined as the ratio $C_i/C_0$. Generally contrast efficiency is improved in emission computed tomography because the effects due to overlying or underlying structures are reduced.

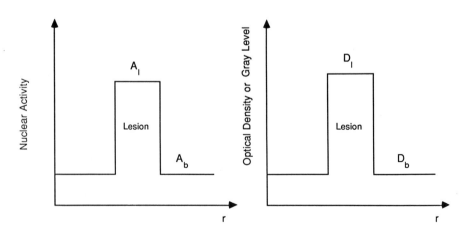

**Figure 169**  Radioactivity distribution of a phantom.

## C. Image Noise

Image noise can be characterized by magnitude and texture. Noise magnitude in radionuclide imaging is caused by random noise fluctuations from the statistical nature of radiation detection. It can be reduced by increasing detected counts or by smoothing techniques. Noise texture is determined by the recording device and image processing techniques used. For example, a recording device having smaller density dots produces "sharper" noise texture. Smoothing filters tend to blur out noise structures producing "broader" noise texture. Both the magnitude and texture of radionuclide image noise have important effects on lesion detection and clinical diagnosis.

# VII. Internal Radiation Dosimetry and Biological Effects

Radiation emissions from radioactivity decay such as alpha, beta, and gamma radiations all carry sufficiently high energy to ionize atoms or molecules and are called *ionizing radiations*. As discussed in Section VI of Chapter 1, ionizing radiations are capable of producing deleterious effects to the body in high enough doses. In nuclear medicine, the unit *gray* is often used to measure the radiation dose absorbed by a tissue. It is defined as 1 joule of energy absorbed by 1 kg of tissue, and 1 gray = 100 rads. Another unit also frequently used is the *dose equivalent*, which takes into consideration the ionization power of different types of ionizing radiation. The unit for the dose equivalent is *rem*, which is related to rad by the following equation:

$$\text{rem} = \text{QF} \times \text{rad} \tag{3-43}$$

where QF is the *quality factor* of the radiation indicating its ionization power. The quality factors for different types of radiation are listed in Table XI.

Calculation of the radiation dose received by a patient in a nuclear medicine procedure requires information about the biodistribution of the radionuclide in various tissue organs, the types and energies of the radiation emissions, physical half-life, biological half-life, and the sizes and relative positions of the source and target organs. This calculation is conventionally done based on a standard man (weight = 70 kg, height = 172 cm, surface area = 1.85 $cm^2$, etc). The organ that receives the most radiation dose is called the *critical organ*. Specifically,

**Table XI**
Quality factors for various types of radiation

| Radiation | QF |
|-----------|-----|
| $\alpha$ | 20 |
| $\beta$ | 1 |
| $\gamma$ | 1 |
| neutron | 10 |
| proton | 10 |
| X-ray | 1 |

it is important to know the radiation dose received by the gonads, or the *gonad dose,* in a particular procedure. Table XII lists the critical organ dose and gonad dose for a number of clinical procedures.

In the following, an example is given to illustrate the absorbed dose calculation. In this example, the idea that radioisotopes with short half-lives give less patient exposure will become clear. First, let us assume that a radioisotope is deposited in an organ at time $t = 0$ with little biological excretion; that is, the biological half-life is much greater than the physical half-life. The total number of radioactive nuclei deposited is $N_0$ and the radionuclide decays according to

$$N(t) = N_0 e^{-\lambda t} \qquad (3\text{-}44)$$

where $\lambda$ is the decay constant. The nuclear activity $A(t)$ at time $t$ is given by

$$A(t) = -\frac{dN}{dt} = N_0 \lambda e^{-\lambda t} = \lambda N(t) \qquad (3\text{-}45)$$

where $A(0) = \lambda N_0$. The quantity $A$ has the unit *becquerel* (one disintegration or transformation per second) or *curie*. The cumulated activity $A'$ between time $t_1$ and $t_2$ is

$$A'(t_1, t_2) = \int_{t_1}^{t_2} A(t) \cdot dt = A(0) \cdot \frac{(e^{-\lambda t_1} - e^{-\lambda t_2})}{\lambda} \qquad (3\text{-}46)$$

The total cumulated activity is given by setting $t_1 = 0$, and $t_2 = \infty$, that is,

$$A'(0, \infty) = \frac{A(0)}{\lambda} = N_0 \qquad (3\text{-}47)$$

or

$$A'(0, \infty) = 1.443 A(0) T_{1/2} \qquad (3\text{-}48)$$

Equation (3-47) shows that the total cumulated activity $A'(0, \infty)$ is equal to the total number of radionuclide $N_0$ at time $t = 0$. Alternatively, $A'(0, \infty)$ can be expressed as a product of the activity at time $t = 0$ and the half-life of the radionuclide. This implies that given two radionuclides having the same initial activity, the total cumulated activity or radiation dose to the patient differs in proportion to the half-lives of the radionuclides. If the biological half-life of the radioiso-

**Table XII**
Critical organ and gonad doses for several important radiopharmaceuticals

| Radiopharmaceutical | Procedure | Critical organ | Critical organ dose (mGy/mCi) | Gonad dose (mGy/mCi) |
| --- | --- | --- | --- | --- |
| $^{99m}$Tc pertechnetate | brain scan | intestine | 1.3 | 0.27 |
| $^{99m}$Tc pertechnetate | thyroid scan | intestine | 2.5 | 0.2 |
| $^{99m}$Tc microspheres | lung scan | lung | 2.1 | 0.05 |
| $^{99m}$Tc pyrophosphate | bone scan | bladder | 4.0 | 0.27 |
| $^{131}$I hippuric acid | kidney scan | bladder | 100.0 | 1.0 |

tope, that is, the rate at which the radionuclide is excreted by the body, is comparable to its physical half-life, then the half-life $T_{1/2}$ in Eq. (3-48) should be replaced by $T_{1/2}^{eff}$ given by Eq. (3-11).

Suppose that 1 mCi of $^{99m}$Tc labeled microspheres is inhaled by a patient for lung scan. Assuming that the lung has a mass of 1 kg, the fraction received by the lung per unit activity is 0.0421, and the biological half-life of the radio-labeled microspheres is much longer than the physical half-life, the radiation dose per unit mass received by the lung, $E_{rad}$, can be calculated using Eq. (3-48), that is,

$$E_{rad} = A'(0, \infty)$$
$$\times \frac{\text{fraction of radiated energy received by the lung per unit of activity}}{\text{mass of the lung}}$$
$$= (1.443 \times 3.7 \times 10^7)[\sec^{-1}] \times (6 \times 60 \times 60)[\sec]$$
$$\times \frac{0.0421 \times 1.4 \times 10^5 \times 1.6 \times 10^{-19}[\text{joule}]}{1.0[\text{kg}]}$$
$$= 1.03 \times 10^{-2}[\text{gray}]$$

## Problems

1. A sample of $^{99m}$Tc containing 4 mCi/mL was prepared at 10 A.M. If 1 mCi will be needed to be injected into a patient at 3 P.M., what volume should be withdrawn from the sample?

2. Find the carrier-free specific activity of $^{99m}$Tc.

3. If a radioactive sample contains a parent–daughter pair, derive the equations for the activity of the daughter in terms of the activities for the parent $A_p$ and daughter $A_d$ at $t = 0$ and the decay constants, $\lambda_p$ and $\lambda_d$.

4. The biological half-life of iodine in the thyroid is 25 days. $^{132}$I and $^{125}$I have respectively physical half-lives of 2.3 hours and 60 days. Calculate their effective half-lives.

5. A parent radionuclide A decays to a daughter radionuclide B with decay constant $\lambda_A$. The daughter B then decays to a stable nuclide C, with decay constant $\lambda_B$. Initially, at $t = 0$, we have $N^A = N_0^A$ and $N^B = N^C = 0$. (a) Starting with the basic differential equations, derive expressions for $N^A(t)$, $N^B(t)$, $N^C(t)$. (b) What approximations can be made if $\lambda_A \gg \lambda_B$ and if $\lambda_A \ll \lambda_B$? (c) Sketch $N^A$, $N^B$, and $N^C$ versus $t$ for $\lambda_A \gg \lambda_B$, $\lambda_A = \lambda_B$, and $\lambda_A \ll \lambda_B$.

6. What is the best time to elute a Mo → Tc generator for maximum amount of $^{99m}$Tc activity?

7. The attenuation coefficient for gamma ray at 140 keV in water is 0.151 cm$^{-1}$. Calculate the half-value layer thickness.

8. What is the blood volume of a patient if 5 mL of $^{123}$I-labeled albumin with a net count rate of $1.5 \times 10^5$ counts/sec was injected into the blood and the net count rate of a 5-mL sample drawn 5 hours later was $10^2$ counts/sec?

9. A gamma camera is used to image a phantom in air as shown. The phantom is filled with $^{99m}$Tc and contains a spherical void filled with air. Draw the image and discuss how geometrical variations can affect detection of the spherical void.

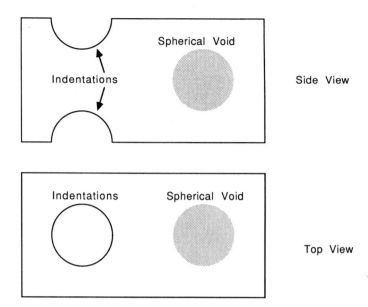

10. List the factors and explain how they affect the spatial resolution, detection efficiency, and deadtime of a scintillation camera system used in nuclear medicine imaging.

11. Suppose two radioisotope X and X' have half-lives $T_{1/2}$ and $T'_{1/2}$, respectively, and $T_{1/2} < T'_{1/2}$. Both radioisotopes emit one particle/nucleus at 0.140 keV. The atomic weights of the two isotopes are $A$ and $A'$. If one patient is given $A$ grams of X and the other $A'$ grams of X', which patient will receive more exposure? If one patient is given 1 mCi of X and the other 1 mCi of X', which patient will receive more exposure? Assume the biological half-life is $\gg T_{1/2}$ or $T'_{1/2}$.

12. Derive the equation for the geometric mean of counts from two point sources located along a line through the axis of two opposing detectors. Compare it to Eq. (3-38) for $t = 0$.

## References and Further Reading

Anger, H. O. (1958). *Rev. Sci. Instr.* **29,** 27.

Anger, H. O. (1964). *J. Nucl. Med.* **5,** 515.

Anger, H. O., Price, D. C. and King, P. H. (1967). *J. Nucl. Med.* **8,** 314.

Anger, H. O. (1971). "Tomographic Gamma-ray Scanner with Simultaneous Readout of Several Planes" *In* "Fundamental Problems in Scanning" (A. Gottschalk and R. N. Beck, eds.) C. C. Thomas, pp. 195–211. Springfield, Illinois.

Blahd, W. H. (1965). "Nuclear Medicine, 2nd ed." McGraw-Hill, New York.

Brownell, G., Burnham, C., Correia, J., Chesler, D., Ackerman, R., and Tavares, J. (1979). *IEEE Trans. Nucl. Sci.* **NS-26,** 2698.

Budinger, T. F., and Gullberg, G. T. (1974). *IEEE Trans. Nucl. Sci.* **NS-21,** 2.

Budinger, T. F., Gullberg, G. T., and Huesman, R. H. (1979). Emission computed tomography. *In* "Image Reconstruction from Projections: Implementation and Applications". (G. T. Herman, ed) pp. 147–246. Springer-Verlag, New York.

Casey, M. E., and Nutt, R. (1986). *IEEE Trans. Nucl. Sci.* **NS-33,** 460.

Cassen, B., Curtis, L., Reed, C., and Libby, R. (1951). *Nucleonics* **9,** 46.

Cassen, B. (1965). *J. Nucl. Med.* **6,** 767.

Chang, L. T. (1978). *IEEE Trans. Nucl. Sci.* **NS-25,** 638.

Cho, Z. H., Chan, J. K., and Eriksson L. (1976). *IEEE Trans. Nucl. Sci.* **NS-23,** 613.

Cho, Z. H., Farukhi, M. R. (1977). *J. Nucl. Med.* **18,** 840.

Derenzo, S. E., Budinger, T. F., Cahoon, J. L., Huesman, R. H., and Jackson, H. G. (1979). *IEEE Trans. Nucl. Sci.* **NS-26,** 2790.

Eriksson, L., Bohm, C., Kesselberg, M., Holte, S., Bergström, M., and Litton, J. (1987). *IEEE Trans. Nucl. Sci.* **NS-34.** 344.

Gullberg, G. T., and Budinger, T. F. (1981). *IEEE Trans. Biomed. Eng.* **BME-28,** 142.

Harper, P. V., and Beck, R. N. (1965). *J. Nucl. Med.* **6,** 332.

Hine G. J., and Sorenson, J. A. (1974). "Instrumentation of Nuclear Medicine, Vol. 2." Academic Press, New York.

Jaszczak, R. J., Huard, D., and Murphy, P. (1976). *J. Nucl. Med.* **17,** 511.

Jaszczak, R. J., Murphy, P. H., Huard, D., and Burdine, J. A. (1977). *J. Nucl. Med.* **18,** 373.

Jaszczak, R. J., Chang L. T., and Murphy, P. H. (1979). *IEEE Trans. Nucl. Sci.* **NS-26,** 610.

Jaszczak, R. J., Coleman, R. E., and Lim, C. B. (1980). *IEEE Trans. Nucl. Sci.* **NS-27,** 1137.

Jaszczak, R. J., Greer, K. L., and Coleman, R. E. (1988). *J. Nucl. Med.* **29,** 1398.

Jaszczak, R. J. (1988). *IEEE Proc.* **76,** 1079.

Kanno, I., Uemura, K., Miura, S., and Miura, Y. (1981). *J. Comput. Assist. Tomogr.* **5,** 216.

Keyes, J. W., Orlandea, N., Heetdesks, W. J., Leonard, P. F., and Rogers, W. L. (1977). *J. Nucl. Med.* **18,** 381.

Kuhl, D. E., and Edwards, R. Q. (1963). *Radiology* **80,** 653.

Kuhl, D. E., Edwards, R. Q., Ricci, A. R., Yacob, R. J., Mich, T. J., and Alavi, A. (1976). *Radiology* **121,** 405.

Laval, M., Moszyński, M., Allemand, R., Cormoreche, E., Guinet, P., Ordu, R., and Vacher, J. (1983). *Nucl. Instr. Meth.* **206,** 169.

Llacer, J. (1981). *IEEE Spectrum* July, 33.

Malko, J. A., Van Heertum, R. L., Gullberg, G. T., and Kowalsky, W. P. (1986). *J. Nucl. Med.* **27,** 701.

Metz, C. E., Tsui, B. M. W., and Beck, R. N. (1974). *J. Nucl. Med.* **15,** 1078.

Metz, C. E., and Doi, K. (1979). *Phys. Med. Biol.* **24,** 1079.

Metz, C. E. (1980). *Phys. Med. Biol.* **25,** 1059.

Muehllehner, G. (1971). *Phys. Med. Biol.* **16,** 87.

National Electrical Manufacturers Association (1980). "Performance Measurements of Scintillation Cameras." Standards Publications/No. Nu-1.

Nudelman, S., and Patton, D. D. (1980). "Imaging for Medicine Vol. 1, Nuclear Medicine, Ultrasonics and Thermography." Plenum, New York.

Patton, J. A., Brill, A. B., and King, P. H. (1973). Transverse section brain scanning with a multicrystal cylindrical imaging device. *In* "Tomographic Imaging in Nuclear Medicine" (G. S. Freeman, ed.) Soc. Nucl. Med., pp. 28–43. New York.

Phelps, M. E., and Mazziotta, J. C. (1985). *Sci.* **228,** 799.

Rankowitz, S. (1962). *IRE Int. Con. Rec.* **10,** 49.

Robertson, J. S., Marr, R. B., Rosenblum, M., Radeka, V., and Yamamoto, Y. L. (1973). Thirty-two-crystal positron transverse section detector. *In* "Tomographic Imaging in Nuclear Medicine," pp. 142–153. Society of Nuclear Medicine, New York.

Rollo, D. F. (1977). "Nuclear Medicine: Physics, Instrumentation and Agents." Mosby, St. Louis.

Sorenson, J. A. (1974). Quantitative measurement of radioactivity *in vivo* by whole-body counting. *In* "Instrumentation of Nuclear Medicine", Vol 2 (G. J. Hine, and J. A. Sorenson, eds), pp. 311–347. Academic Press, New York.

Sorenson, J. A., and Phelps, M. E. (1987). "Physics in Nuclear Medicine, 2nd ed." Grune & Stratton, Orlando, Florida.

Stokeley, E. M., Sweinsdottir, E., Lassen, N. A., and Rommer, P. J. (1980) *J. Comp. Assist. Tomogr.* **4**, 230.

Tanaka, E. (1987). *IEEE Trans. Nucl. Sic.* **NS-34**, 313.

Ter-Pogossian, M. M., Phelps, M. E., Hoffman, E. J., and Mullani, N. (1975). *Radiology* **114**, 89.

Ter-Pogossian, M. M., Mullani, N. A., Hood, J., Higgins, C. S., and Curie, M. (1978). *Radiology* **128**, 477.

Ter-Pogossian, M. M., Mullani, N. A., Fiscke, D. C., Markham, J., and Snyder, D. L. (1981). *J. Comput. Assist. Tomogr.* **5**, 227.

Thompson, C. J., Yamamoto, Y. L., and Meyer E. (1978). *J. Comput. Assist. Tomogr.* **2**, 649.

Tretiak, O. J., and Metz C. E. (1980). *SIAM J. App. Math.* **39**, 341.

Tsui, B. M. W., Chen, C. T., Yasillo, N. J., Ortega, C. J., Charleston, D. B., Harper, P. V., and Lathrop, K. A. (1981). *Proceedings of the Third International Radiopharmaceutical Dosimetry Symposium,* HHS Publication FDA 81-8166, pp. 138–156.

Tsui, B. M. W., Gullberg, G. T., Edgerton, E. R., Gilland, D. R., Perry, J. R., and McCartney, W. H. (1986). *J. Nucl. Med.* **27**, 810.

Tsui, B. M. W. (1988). "Collimator Design, Properties, and Characteristics." *In* "The Scintillation Camera", (G. H. Simmons, ed.) pp. 17–45. The Society of Nuclear Medicine, New York.

Tsui, B. M. W., Gullberg, G. T., Edgerton, E. R., Ballard, J. G., Perry, J. R., McCartney, W. H., and Berg, J. (1989). *J. Nucl. Med.* **30**, 497.

Webb, S. (1988). "The Physics of Medical Imaging." Adam Hilger, Bristol and Philadelphia.

Williams, C. W., Crabtree, M. C., and Burgiss, S. G. (1979). *IEEE Trans. Nucl. Sci.* **NS-26**, 619.

Williams, J. J., Snapp, W. P., and Knoll, G. F. (1979). *IEEE Trans. Nucl. Sci.* **NS-26**, 628.

Wrenn, F. R., Good, M. L., and Handler, P. (1951). *Sci.* **113**, 525.

# CHAPTER 4

# Magnetic Resonance Imaging

The phenomenon of nuclear magnetic resonance (NMR) discovered by F. Block and E. Purcell in 1946 has become a standard spectroscopic technique in chemistry and physics. For this discovery, Block and Purcell were awarded the Nobel Prize in 1952. More recently NMR has been applied as an imaging technique pioneered by P. Lauterbur (1973), P. Mansfield (1973), and R. Damadian (1971). Nuclear magnetic resonance imaging, abbreviated MRI in the clinical field, has the advantages of being able to penetrate bony and air-filled structures with negligible attenuation and artifact. The modality uses non-ionizing radiation and is minimally invasive. It is capable of providing excellent soft-tissue contrast with imaging in any arbitrary plane. Although formerly used primarily for examination of the brain and spinal cord, the development of rapid imaging techniques has extended the role of NMR imaging to the chest and abdomen where motion had previously been a limitation. The recent development of flow imaging provides the possibility of performing NMR angiography. Moreover, with the use of spectroscopy and functional imaging it is capable of yielding information about the physiological state of the tissue.

## I. Fundamentals of Nuclear Magnetic Resonance

Like many forms of spectroscopy NMR utilizes electromagnetic radiation to probe the fundamental properties of matter. In NMR the radiation employed is in the radio-frequency (RF) portion of the electromagnetic spectrum, and the property studied is the interaction of this radiation with the nuclear magnetic moment. An understanding of the principles of nuclear magnetic resonance imaging is founded on principles of quantum and classical mechanics.

## A.  Angular Momentum

An object with a mass, $m$ orbiting about an axis with velocity of **v**, as shown in Fig. 170, has an angular momentum $\mathbf{L_a}$.

$$\mathbf{L_a} = m\mathbf{r} \times \mathbf{v} \qquad (4\text{-}1)$$

where $\mathbf{L_a}$, **r**, and **v** are all vectors. A spinning object also possesses an angular momentum whose vector points in the direction of the thumb when the right-hand rule is applied (Fig. 171). Spinning is a natural phenomenon for electrons and nuclear particles. Besides orbiting about the nuclear axis, electrons, protons, and neutrons spin about their own axes. As a result of these motions there are two angular momenta associated with each electron and nuclear particle: orbital angular momentum and spin angular momentum.

In contrast to macroscopic particles, electrons and nuclear particles obey the rules dictated by quantum mechanics. These rules account for the wave properties of matter, and limit the possible energy of angular momentum to discrete values. Physical properties such as angular momentum are said to be quantized—a phenomenon that cannot be explained on the basis of classical mechanics. The magnitude of the angular momentum of a subatomic particle observed about an arbitrary $Z$ axis is limited to the values:

$$L_{az} = m_I(h/2\pi) \qquad (4\text{-}2)$$

where $h$ is the Planck constant equal to $6.6 \times 10^{-34}$ joule-sec, and $m_I$ is the magnetic quantum number. The combination of all nuclear angular momenta in a nucleus generates a single number $I$, called the spin quantum number and determined by the spin of unpaired neutrons or protons. For a given nucleus the value of $I$ must be 0, $\pm(1/2)n$, where $n$ is an integer. When the nucleus has an odd atomic number or an odd number of neutrons or both, $I$ is nonzero, resulting in a net nuclear angular momentum. Nuclei with even numbers of protons and neutrons have no net spin ($I = 0$). It will become clear later that these nuclei do not possess a nuclear magnetic moment, and therefore cannot be imaged by MRI. The magnetic quantum number that describes the eigenstates or energy levels of the nucleus is related to the spin quantum number as

$$m_I = I, I - 1, I - 2, \ldots, -I$$

Note that the total number of energy levels possible is equal to $2I + 1$. For the case of the proton $I = 1/2$, and the $z$-component of the angular momentum, of-

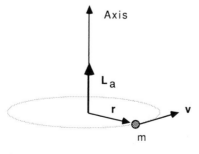

**Figure 170**   Object with mass $m$ orbiting around a designated axis at a radius **r** and with velocity **v** has an angular momentum $\mathbf{L_a}$.

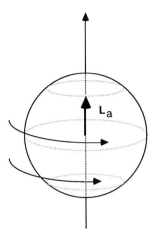

**Figure 171**   Angular momentum of a spinning object.

ten termed the nuclear "spin," can only exist in one of two energy states: $m_I = +1/2$ or $m_I = -1/2$.

By convention the angular momentum is depicted by a vector perpendicular to the plane of rotation. Particles with a counterclockwise rotation in the $XY$ plane have a positive spin angular momentum in which the vector is aligned along the positive $Z$ axis, while those with negative momentum undergo clockwise rotation. A pair of particles rotating in opposite directions is in the lowest energy state, with a net angular momentum of zero.

## B.  Magnetic Dipole Moment

Due to their charge properties all nucleons including neutrons have a magnetic dipole moment. It is well known that electrons and protons possess the same charge but different sign. It is less well known, however, that although the net charge is zero, neutrons have an asymmetrical charge distribution in the particle. A unit called the nuclear magneton $\mu_{mN}$ is used to express the magnitude of the magnetic dipole moment of the nucleus.

$$1 \; \mu_{mN} = 5.05 \times 10^{-27} \, J/tesla$$

The magnetic dipole moments of the proton and the neutron are respectively $\mu_{mp} = 2.79 \; \mu_{mN}$ and $\mu_{mn} = -1.91 \; \mu_{mN}$ where the + or − sign indicates whether the magnetic dipole moment is in or opposite to the direction of angular momentum.

The nuclear magnetic moment $(\mu_m)$ is related to the nuclear angular momentum through the expression

$$\boldsymbol{\mu}_m = \gamma \mathbf{L}_{az} \tag{4-3}$$

where $\gamma$, termed the gyromagnetic ratio, is a constant characteristic of each particular nucleus. Magnetic dipole moments of a few important nuclei can be found in Table XIII. $^1H$, $^2H$, $^7Li$, $^{13}C$, $^{14}N$, $^{23}Na$, $^{31}P$, and $^{127}I$ are just a few examples of the nuclei having a net spin and a magnetic dipole moment. The hydrogen nucleus, a single proton, is of particular importance in NMR imaging because of its

## Table XIII
Nuclear properties of a few important nuclei in magnetic resonance imaging

| Nucleus type | Mag. dipole moment, $\mu_{mN}$ | Nuclear spin number, I | Gyromagnetic ratio (radians/tesla) | Larmor freq. (MHz/tesla) |
|---|---|---|---|---|
| $^1$H | 2.79 | 1/2 | $2.7 \times 10^8$ | 42.6 |
| $^2$H | 0.85 | 1 | $4.1 \times 10^7$ | 6.5 |
| $^{13}$C | 0.70 | 1/2 | $6.7 \times 10^7$ | 10.7 |
| $^{14}$N | 0.40 | 1 | $1.9 \times 10^7$ | 3.1 |
| $^{23}$Na | 2.21 | 3/2 | $7.1 \times 10^7$ | 11.7 |
| $^{31}$P | 1.13 | 1/2 | $1.1 \times 10^8$ | 17.2 |

abundance in biological tissues. Currently all clinical images are obtained using the proton signal.

As with angular momentum the nuclear magnetic moment is quantized and can exist only in discrete energy levels described by

$$\mu_m = \gamma m_I (h/2\pi) \qquad (4\text{-}4)$$

For the proton, with a spin $= 1/2$, two energy levels are allowed for the nuclear magnetic moment. This is graphically illustrated in Fig. 172. In the absence of an applied magnetic field the energies of the two eigenstates are identical and they are said to be degenerate. It is impossible to observe transitions between degenerate states, and therefore an NMR signal is not observed. However, when an external magnetic field $B_0$ is applied, the interaction $\mu_m$ with $B_0$ produces a difference in the energy ($\Delta E$) of the two states. The energy levels differ according to the following equation

$$\Delta E = 2\mu_m B_0 \qquad (4\text{-}5)$$

As demonstrated in Fig. 172, the energy difference of the eigenstates is directly proportional to the applied magnetic field. The loss of degeneracy in the presence

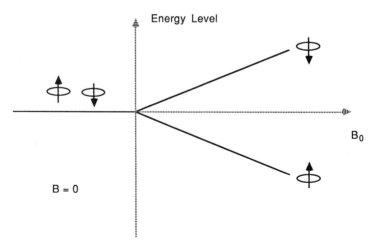

**Figure 172** Energy diagram for spin $= 1/2$ nucleus. In the absence of an external magnetic field ($B = 0$) the energy levels of the nuclear spin states are identical and the states are degenerate. The application of a magnetic field produces a difference in energy of the two spin states through Zeeman splitting.

of an applied magnetic field is known as the Zeeman effect and forms the fundamental basis for magnetic resonance imaging, as we shall see in Section III.

## C. Magnetization

For protons in an external magnetic field, the individual nuclear magnetic moments align themselves in one of two energy states. The lower energy state, in which the magnetic moment is aligned parallel to the magnetic field, corresponds to $m_l = +1/2$, and is the ground state. Likewise nuclei with their magnetic moment aligned antiparallel to the applied magnetic field reside in a higher energy state where $m_l = -1/2$. The relative population difference between the two energy levels is governed by the Boltzmann distribution.

$$n_{\text{upper}}/n_{\text{lower}} = \exp(-\Delta E/k_b T) \tag{4-6}$$

where $n_{\text{upper}}$ and $n_{\text{lower}}$ are populations of nuclei in the higher energy and ground states, respectively, $k_b$ is the Boltzmann constant equal to $1.38 \times 10^{-23}$ joules/K, and $T$ is the absolute temperature. To decrease the overall energy of the spin system the nuclei preferentially occupy the ground state that has lower energy. This process is limited by the effects of thermal motion, which act to randomize the population levels. For physiologic temperatures and at $B_0$ field strengths typically employed for imaging, the difference in population is approximately 1 part per million. That is, for every million nuclei residing in the higher energy state, 1 million and 1 nuclei are present in the lower energy state.

The small difference in population described by the Boltzmann distribution places fundamental limitations on the NMR experiment. First, the signal detected and processed is only due to the small net difference in the population of the energy levels. Thus NMR is an inherently insensitive technique compared to other forms of electromagnetic spectroscopy where the population differences are orders of magnitude greater. Second, the population difference calculated from Eq. (4-6), and hence the NMR signal, can be increased either by increasing the energy difference between states or by reducing the temperature of the system. Since humans respond poorly to subfreezing temperatures, one is limited to increasing the energy difference between states in order to increase the NMR signal. This can be accomplished by increasing the magnitude of the applied magnetic field. Thus in theory, greater signal is possible by using stronger applied magnetic fields.

Thus far we have described the properties of angular momentum and nuclear magnetic moment using the model of quantum mechanics. This model is necessary to explain the quantized nature of these properties. However, NMR spectroscopy is somewhat unique in that many experiments can be accurately explained using concepts of classical physics. This is possible because the predictions obtained using classical and quantum mechanics converge when $\Delta E$ is much less than $k_b T$, as is true for NMR. The choice of model to describe the NMR experiment can be made based on simplicity without a great loss in accuracy. For the imaging experiment the classical model, in which magnetic vectors are used to describe the nuclear magnetic moments, provides a simple and convenient description of the physical processes.

In the classical model the nuclear magnetic moment is represented by a small magnetic vector. If this vector is placed in a static magnetic field $B_0$, the force tends to pull it to the direction of the applied field just like a small bar magnet, as shown in Fig. 173. Because of its spinning motion, the nucleus responds to this

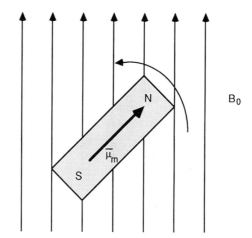

**Figure 173** Magnetic vector placed in an external magnetic field.

force by undergoing a complex motion known as precession. Precessional motion is similar to that experienced by a gyroscope. For the gyroscope to precess, two events must occur. First, it has to possess an angular momentum. Second, a torque or force must exist to tilt it. The angular momentum is provided by the spinning motion. The force is provided by gravity. These mechanisms are available for nuclear particles as well. Nuclear spin provides the angular momentum. The interaction of the external magnetic field and the nuclear magnetic dipole moment provides a torque to start the nuclear precession. The NMR sample can be described as a collection of microscopic dipoles undergoing precession about the axis of the applied magnetic field. By convention this axis is placed in the positive $Z$ direction. The small difference in the population of energy levels produces a small net magnetization termed $M_z$, oriented along the positive $Z$ axis (Fig. 174). Because the individual nuclei precess about the $Z$ axis in a random fashion, the transverse or $XY$ component of the nuclear magnetization cancels to zero. Since $M_z$ represents the vector sum of the individual magnetic moments, it does not precess. However, an external torque can be applied to tilt $M_z$ away from the $Z$ axis. This action generates a vector component that rotates in the $XY$ plane. As we shall see, this torque can be generated by the application of an RF magnetic field.

## D. Larmor Frequency

At equilibrium the populations of the energy states described in the quantum mechanical model are stable. For every transition from the ground state to the higher energy state there is a transition occurring from the higher energy state to the ground state. Thus, the net exchange of energy between the spin system and the outside world is zero. If the spin system is irradiated with electromagnetic irradiation of frequency $f_0$ where

$$\Delta E = hf_0 = \gamma(h/2\pi)B_0 \qquad (4\text{-}7)$$

transitions will be induced between energy levels. This is the resonance condi-

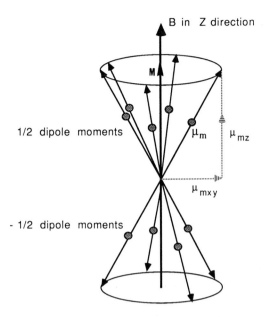

B in Z direction

**Figure 174**  Precessing protons produce a small net magnetization vector M which is aligned parallel to the axis of the applied magnetic field. Due to the random distribution of the individual spins about the Z axis, the net magnetization vector in the XY plane is zero.

tion. Irradiation with a frequency other than $f_0$ will not interact with the spin system to stimulate such transitions. The probability of inducing a transition between energy levels is proportional to the population of the energy state. Since at equilibrium the ground state is preferentially populated, irradiating the system with radio-frequency energy at the resonance frequency increases the relative excited state population. If the frequency is expressed in radians per second ($\omega = 2\pi\nu$) this equation can be simplified to

$$\omega_l = \gamma B_0 \qquad (4\text{-}8)$$

This fundamental expression is known as the Larmor equation and $\omega_l$ is known as the Larmor or resonant frequency. This equation reveals how the resonant frequency of a nuclear spin is directly proportional to the applied magnetic field. As we will discuss in Section III of this chapter, this forms a basic principle for obtaining spatial information from the NMR signal.

The gyromagnetic ratio is the proportionality constant between the Larmor frequency and the applied magnetic field. Since a nucleus of different atomic structure has a different $\gamma$, it will resonate at a different frequency when exposed to the same magnetic field. The gyromagnetic ratios of a few important nuclei can be found in Table XIII. Likewise, a given nucleus will have a different resonant frequency when placed in a different magnetic field strength. For a proton, $\gamma = 2.7 \times 10^8$ rad/T and the Larmor frequency is 42.57 MHz at 1 tesla. At a field strength of 10 T the $\Delta E$ of the energy levels increases proportionally, and the protons in the sample resonate at 425.7 MHz. Note that 1 tesla or (T) is equal to $10^4$ gauss (G). For comparison the earth's magnetic field is approximately 0.5 G.

# E.   Rotating Frame of Reference and the RF Magnetic Field

At equilibrium in the classical model, the magnetization vector $M_z$ is aligned along the $Z$ axis. Based on the classical mechanical observation that maximal interaction between two magnetic dipoles occurs when they are perpendicular, it follows that the maximal torque can be applied to $M_z$ if a magnetic field is applied in a perpendicular plane. This magnetic field is the magnetic component of the RF irradiation and is called the $B_1$ field to differentiate it from the $B_0$ or dc magnetic field. On the application of a RF magnetic field at the resonant frequency, $M_z$ is tilted from the $Z$ axis and made to precess at the Larmor frequency in the $XY$ plane as shown in Fig. 175 (a). To simplify the description of this complex motion the rotating frame of reference is employed. In the rotating frame of reference the $X$ and $Y$ axes also rotate about the $Z$ axis at the Larmor frequency. As illustrated in Fig. 175 (b) the precessing magnetization remains stationary relative to the rotating frame of reference used to describe its motion.

NMR techniques utilize RF pulses to selectively change the orientation of the magnetization vector. The duration of the RF pulse or pulse width (PW) modulates the RF carrier frequency to produce an excitation bandwidth of approximately 1/PW. Thus a short RF pulse of 1 $\mu$sec would excite nuclei resonating over a frequency bandwidth of $10^3$kHz centered around the carrier frequency. These short-duration pulses are called hard or nonselective pulses, since they are designed to produce excitation over a broad frequency range. An RF pulse of duration 2 msec would produce excitation over a narrow frequency bandwith of 500 Hz. These long-duration pulses are called soft or selective pulses because they can be used to selectively excite nuclei residing in a narrow frequency range. The frequency selectivity of the soft pulse makes them useful tools in MRI. For example, a soft pulse is used in combination with a linear magnetic field gradient to selectively excite a slice of nuclei within a larger sample. In this case the gradient produces a linear dispersion of resonant frequencies over a range of several kilohertz. The simultaneous application of a selective pulse, with a frequency bandwidth of several hertz, excites only those nuclei in a narrow slice. The slice thickness is a function of the slope of the field gradient and the frequency bandwidth of the soft pulse.

The ultimate position of the magnetization vector following the application of the RF magnetic field depends on the amplitude of the $B_1$ field and the duration of exposure. The angle traced by the magnetization vector following application of the $B_1$ field is termed the flip angle. The strength of the $B_1$ field employed in NMR experiments is often described in terms of flip angle. For example, a 90° pulse would describe a $B_1$ field of sufficient magnitude and duration to tilt the $M_z$ vector 90° into the $XY$ plane, whereas a 180° pulse would tip $M_z$ onto the negative $Z$ axis. In the case of hard or nonselective RF pulses the PW must be kept short in order to produce uniform excitation over a large frequency bandwidth. For these pulses the maximal RF amplitude is usually employed, and the strength of the $B_1$ field needed to obtain the desired flip angle is adjusted by varying the PW over a narrow range. In the case of imaging, the PW must be kept constant since it directly controls the bandwidth and therefore the thickness of the selected slice. For these studies the strength of the $B_1$ field is adjusted by varying the amplitude of the RF field while maintaining a constant PW.

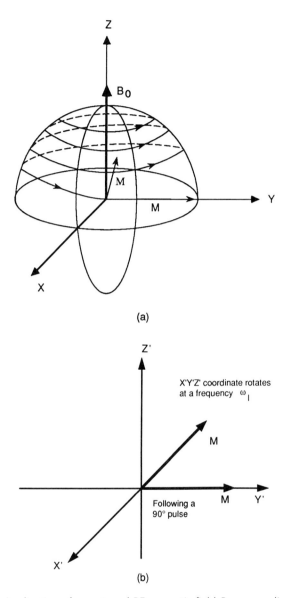

(a)

(b)

**Figure 175**   Application of an external RF magnetic field $B_1$, perpendicular to the dc magnetic field $B_0$, tips the magnetization away from the $Z$ axis toward the $XY$ plane. In the laboratory frame of reference (a) the nutated magnetization precesses about the axes of both the $B_0$ and $B_1$ magnetic fields. The motion is represented by a spiral path from the $Z$ axis to the $XY$ plane. In the rotating frame of reference (b) the coordinate system rotates about the $Z$ axis at the Larmor frequency. In this representation, the nutated magnetization is represented by a simple rotation onto the $XY$ plane.

## F.   Free Induction Decay (FID)

A basic system for detecting the NMR signal is shown in Fig. 176. A small sample of material is placed in the static magnetic field, $B_0$. The RF transmitter gen-

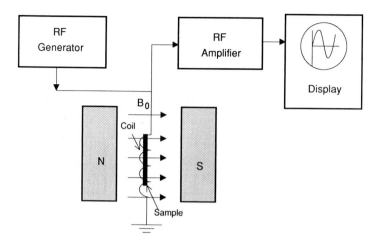

**Figure 176** Simplified schematic for generation and detection of the NMR signal.

erates the $B_1$ magnetic field perpendicular to the static field through a coil surrounding the sample. The $B_1$ field tips the nuclear magnetization vector into the $XY$ plane. After the RF field is turned off, the precessing magnetization vector induces a voltage in the same coil, which is amplified and displayed on the oscilloscope. This signal, illustrated in Fig. 177, is called the free induction decay or FID. It is free because it occurs without the external $B_1$ field. The signal is induced in the receiver coil by the precessing magnetization vector as it decays back to equilibrium. The reason the FID signal is sinusoidal can be understood by examining Fig. 178. The coil responds only to the component of the magnetic field perpendicular to the coil axis. For a coil placed along the $Y$ axis, the $X$-component of $M_{xy}$ rotates around the $Z$ axis inducing a sinusoidal signal.

## G.  Fourier Spectrum of the NMR Signal

Prior to the advent of pulsed techniques the NMR signal was detected by either sweeping the RF frequency or varying the static or dc magnetic field to generate the resonance condition. The signal obtained from these continuous-wave tech-

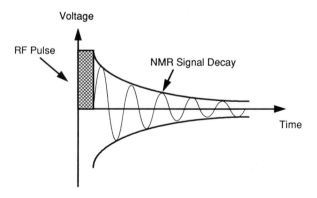

**Figure 177** Free-induction decay signal.

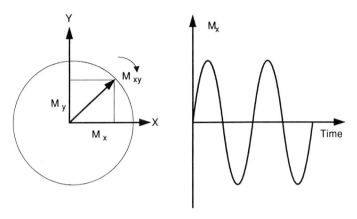

**Figure 178**  Induced voltage in a coil aligned along the Y axis as a function of time.

niques is a spectrum where the amplitude of the signal is plotted as a function of frequency. Continuous-wave techniques have been almost entirely replaced by current pulsed-NMR technology. The major disadvantage of continous-wave methodology is the inherent inefficiency in observing only one frequency at a time. With the advent of pulsed-NMR techniques the entire frequency spectrum is excited simultaneously with a hard RF pulse, and the resulting FID contains all of the information available in the continuous-wave spectrum. Furthermore by adding subsequent FIDs the NMR signal-to-noise ratio can be improved by a factor of $\sqrt{N}$ where $N$ is the number of accumulated scans or FIDs. This occurs because the NMR signal, which is coherent with time, increases by a factor of $N$ as sequential FIDs are added. Since noise is random with time, it increases by a factor of $\sqrt{N}$. The advent of pulsed or Fourier transform NMR techniques allowed NMR technology to be used on less sensitive nuclei such as $^{13}C$ and $^{31}P$, as well as allowing the possibility of NMR imaging.

Although the FID contains the amplitude, phase, and frequency information observed in the NMR spectrum, it is acquired in the time domain, making it more difficult to interpret. The time domain of the FID is related to the frequency domain of the NMR spectrum by a Fourier transformation (Fig. 179). In the NMR experiment sequential FIDs are acquired and stored in the computer until adequate signal-to-noise is obtained. The averaged time domain signal is then

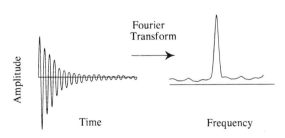

**Figure 179**  Time domain FID signal can be converted into the frequency domain NMR spectrum using a Fourier transform.

converted into an NMR spectrum or frequency domain signal with a Fourier transform for interpretation.

## H. Spin Density

The amplitude of the FID signal shown in Fig. 179 is proportional to the number of spinning nuclei per unit volume defined as spin density of the sample. Therefore, in a system designed to measure water content, NMR signal strength from a sample is determined by the proton concentration of that sample. Typically the water content of soft tissues varies from 50% to 90% (Mansfield and Morris, 1982). The water content of various types of tissues is listed in Table XIV. From this table it can be seen that the contrast based solely on spin density may be rather poor in proton imaging. However, it will become clear later that the MRI signal is dependent on properties of the molecular environment such as the relaxation times $T_1$ and $T_2$. Differences in relaxation times, along with proton density and motion, can be incorporated into the image to enhance contrast among soft tissues.

## I. Relaxation Times

The application of the RF magnetic field introduces energy and order into the nuclear spin system. To return to the original equilibrium state this energy must be lost to the environment through an enthalpy process of spin–lattice relaxation. The order introduced into the system is lost through an entropy process of spin–spin relaxation. In the classical model the induced energy and order of the spin system are represented by the coherent magnetization vector, which is precessing in the $XY$ plane. Suppose that at equilibrium this vector is aligned along the $Z$ axis with a magnitude of $M_0$, and there is no coherent magnetization in the $XY$ plane. Thus two processes must occur for the spin system to return to equilibrium. First the vector along the $Z$ axis must return to its original magnitude ($M_z = M_0$). Second, the magnitude of the vector in the $XY$ plane must return to zero ($M_{xy} = 0$). The rate at which these two processes occur is described using two time constants called simply $T_1$ or the spin–lattice relaxation time, and $T_2$ also known as the spin–spin relaxation time.

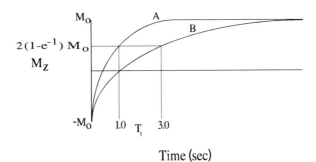

Time (sec)

**Figure 180** Recovery of $M_z$ following a 180° inversion pulse. In a sample with liquid-like properties the $M_z$ recovers quickly and the $T_1$ is relatively short (sample A). A sample with more solid-like properties generally has a longer $T_1$ (sample B). By definition the $T_1$ is the time required for $M_z$ to equal $(1 - e^{-1})M_0$.

**Table XIV**
Water content of fat-free normal
human tissue

| Tissue | Water content (%) |
|---|---|
| Skeletal muscle | 79 |
| Myocardium | 80 |
| Liver | 71 |
| Kidney | 81 |
| Brain white matter | 84 |
| Brain gray matter | 72 |
| Nerve | 56 |
| Femur cortex | 12 |
| Teeth | 10 |

The spin–lattice relaxation time $T_1$ is associated with the return of $M_z$ to $M_0$. In the quantum model of the spin system this process describes the transfer of energy to surrounding structures termed the lattice, as the excited nuclei return to the lower energy ground state. Mathematically the process can be represented with a time constant $T_1$ in the equation:

$$dM_z/dz = (M_0 - M_z)/T_1 \quad \text{or} \quad M_z = M_0(1 - e^{-t/T_1}) \tag{4-9}$$

Since $M_z$ is in the direction of the dc magnetic field, it is difficult if not impossible to measure. Therefore, to measure $M_z$ at an instant of time, it has to be flipped into the $XY$ plane. Special methods such as progressive saturation and inversion recovery are used in the measurement of $T_1$. These methods will be discussed later in this section.

Unlike other forms of electromagnetic spectroscopy, in which the excited and ground states are separated by a large difference in energy, the transition of a nuclear magnetic dipole from the higher energy state to the ground state does not occur through spontaneous emission. For example it takes approximately $10^{-7}$ sec for an excited atom to emit its energy and return to the ground state. In comparison it takes approximately $10^{19}$ sec for an excited nucleus to spontaneously reverse its magnetic moment in a magnetic field of 1 T—a time longer than the estimated age of the universe. In NMR the transfer of energy from the spin system to the lattice occurs through stimulated emission. Small random movements of charged particles generate oscillating magnetic fields. If the frequency of these fields is at the Larmor frequency it can stimulate transitions between the states, allowing the spin system to relax to equilibrium. For protons in a biological system the major mechanism for $T_1$ relaxation is through dipole–dipole interactions. This energy transfer is facilitated through thermal interactions, that is, random collisions between the dipoles and neighboring molecules. The rate of energy transfer between the dipoles and the neighboring structures is dependent on molecular motions. The rapid translational and rotational movement of molecules in a liquid allows this energy redistribution to occur quickly. Therefore the $T_1$ of liquids is shorter than that in solids (Fig. 180). $T_1$ for various biological materials can be found in Table XV (Morgan and Hendee, 1984).

Recall that when the sample is at thermal equilibrium, there is no net transverse magnetization $M_{xy}$. Although there is a net magnetization in the $Z$ direction, the precession around the $Z$ axis of the individual nuclei is out of phase,

**Table XV**
The spin–lattice relaxation time ($T_1$) and spin–spin relaxation time ($T_2$) of various biological tissues at 0.2 tesla

| Tissue | $T_1$, msec | $T_2$, msec |
|---|---|---|
| Fat | 240 ± 20 | 60 ± 10 |
| Muscle | 400 ± 40 | 50 ± 10 |
| Gray matter | 495 ± 85 | 100 ± 10 |
| White matter | 390 ± 70 | 90 ± 20 |
| Lung | 460 ± 90 | 80 ± 30 |
| Kidney | 670 ± 60 | 50 ± 10 |
| Liver | 380 ± 20 | 40 ± 20 |
| Liver metastases | 570 ± 190 | 40 ± 10 |
| Lung carcinoma | 940 ± 460 | 20 ± 10 |

*Source:* Morgan and Hendee, 1984

causing destructive interference in the signal generated by these dipoles. The process in which $M_{xy}$ returns to 0 is characterized by the $T_2$ time constant known as the spin–spin relaxation time. The term spin–spin is used because this process involves the interaction of nuclear spins without the transfer of energy to the surrounding lattice. As a nucleus relaxes from the excited state to the ground state it emits RF energy at the Larmor frequency. This emitted energy can be absorbed by a neighboring nucleus, causing it to be excited to the higher energy level. Thus the spin–spin interaction has the effect of mixing the spin populations, resulting in a loss of signal coherence.

The loss of signal coherence can also occur through mechanisms that are independent of spin–spin interactions. Consider the magnetization generated in the *XY* plane following a 90° RF pulse. If there is a single resonant frequency this magnetization will precess coherently in the *XY* plane. In the rotating frame of reference this is illustrated by a single vector, which remains aligned in the *XY* plane. However, if there are inhomogeneities in the static $B_0$ field there will be a corresponding variation in the resonant frequencies of the nuclear spins exposed to those fields. The nuclei will precess according to the local $B_0$ field that they are currently experiencing. In the rotating frame of reference this is illustrated by a fanning of the spins in the *XY* plane will a subsequent loss of $M_{xy}$. The loss of $M_{xy}$ is described by the term $T_2^*$, which includes the effects of $B_0$ inhomogeneity as well as true spin–spin or $T_2$ relaxation. To measure the true $T_2$, other methods have to be used to remove the effects of $B_0$ inhomogeneity. The spin–echo method, which will be discussed later, is an example. The true decay of $M_{xy}$ devoid of the effect of static dc magnetic field inhomogeneity is governed by the following equation:

$$dM_{xy}/dt = -M_{xy}/T_2 \quad \text{or} \quad M_{xy} = M_0 e^{-t/T_2} \tag{4-10}$$

$T_2$ is always less than or equal to $T_1$ because only molecular motion near the Larmor frequency can affect $T_1$ to cause dipole flipping, but molecular motion at other frequencies can affect $T_2$. As illustrated in Fig. 181, $T_2$ in liquids is usually longer than that in solids. In the case of solids, strong dipole–dipole interaction gives rise to very large $B_0$ field components, which can be one the order of several gauss. In solids it is not unusual to have $T_2$ times of several mseconds rather

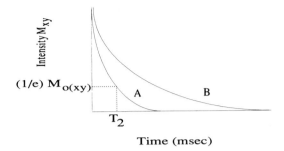

**Figure 181**  Loss of $M_{xy}$ following a 90° RF excitation pulse for two samples with differing $T_2$ times. By definition the $T_2$ is the time required for $M_{xy}$ to decay to $(1/e)M_0$.

than values of several hundred milliseconds typically found in liquids or biological tissue. Because the signal from solids remains coherent for such a short time, special techniques are required to produce spectra and images from these materials. These techniques are currently an active area of research. The $T_2$ values of a few important biological materials are listed in Table XV (Morgan and Hendee, 1984). As seen in this table the $T_1$, $T_2$ times of tissue are altered by many pathologic conditions. The characterization of tissue based on its $T_1$, $T_2$ properties provides useful information in the diagnosis of disease.

## J.  Pulse Sequences

A series of RF pulses and magnetic field gradients can be creatively combined to sensitize the NMR signal to a number of physical properties. Pulse sequences have been used for the measurement of the relaxation times $T_1$ and $T_2$, self-diffusion coefficients, laminar flow, coherent translational motion, and many other physical properties. Many of these spectroscopy experiments have a corresponding imaging experiment in which these physical processes can be described spatially. In this section we will briefly introduce three pulse sequences that are commonly used in spectroscopic studies to measure relaxation times. An understanding of these pulse sequences in necessary to appreciate how contrast, based on the $T_1$ and $T_2$ properties of tissue, is achieved in NMR imaging.

### 1.  Progressive Saturation

Progressive saturation is a commonly used NMR technique for the measurement of $T_1$ relaxation time. It consists of a series of 90° RF pulses separated by an acquisition period in which the resulting FID is sampled, and a delay period ($\tau$). The pulse sequence can be represented simply as $(90° - \text{Acq} - \tau - )^n$. During the acquisition and delay period the magnitude of $M_z$ recovers via spin–lattice relaxation. If the length of $\tau$ is significantly greater than the $T_1$ of the sample, $M_z$ will have returned to original equilibrium magnitude $M_0$ prior to the application of the next RF pulse. If $\tau$ is less than $T_1$ the magnitude of $M_z$ will be less than $M_0$. Following the application of 4 to 5 RF pulses in the series, a new steady-state condition is established in which the energy introduced into the spin system by the RF pulses is balanced by the loss of energy through spin–lattice relaxation. This condition is known as partial saturation. The magnitude of the FID is directly proportional to the magnitude of $M_z$. In the progressive saturation experiment the magnitude of the NMR signal is determined for several values of $\tau$. As demonstrated in Fig.182, signal intensity increases exponentially with the time between RF pulses. This data can be fitted to Eq. (4-9) to calculate $T_1$. As shown in Fig.180 the $T_1$ time is equal to the time necessary for $M_z$ to recover 63% of its equilibrium magnetization.

## 2. Inversion Recovery

The pulse sequence for the inversion recovery experiment consists of an initial 180° pulse, a delay period ($\tau$), and a 90° sampling pulse, which is immediately followed by acquisition of the FID. Following acquisition of the NMR signal an additional delay is introduced to allow $M_z$ to recover to its equilibrium position. As in the progressive saturation experiment, this combination of RF pulses and delays sensitizes the amplitude of the FID to the $T_1$ of the sample. As depicted in Fig. 183, the application of the 180° RF pulse flips the $M_z$ vector onto the $-Z$ axis. In terms of the quantum description this corresponds to a complete inversion of the equilibrium populations of the two energy levels. During the delay period $\tau$ the vector $M_z$ exponentially returns to its equilibrium position along the $+Z$ axis. The 90° pulse is used to sample the magnitude and orientation of $M_z$ by nutating the vector into the $XY$ plane. When phase-sensitive detection is used to collect the FID, the NMR signal is negative, absent, or positive depending on whether the 90° pulse is applied before, during, or after the magnetization has passed through zero. When plotted as a function of $\tau$, the NMR signal increases exponentially from $-M_0$ at $\tau = 0$ to $+M_0$ at $\tau \gg T_1$. Systematic error can arise if the delay following acquisition is not long enough to allow $M_z$ to return fully to $M_0$ before applying the next 180° pulse. Under these conditions a steady-state similar to that generated in the progressive saturation experiment is established, which complicates interpretation of the data. The long delay for full relaxation, usually 3 to 5 times the $T_1$ of the sample, places significant time demands in the inversion recovery study.

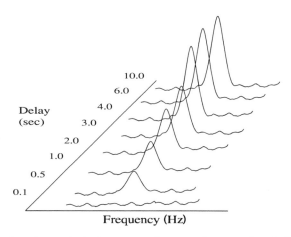

**Figure 182** Signal intensity in the progressive saturation experiment plotted as a function of the time between sequential 90° RF excitation pulses.

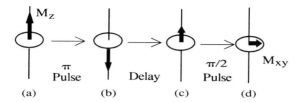

**Figure 183**  Inversion recovery experiment: (a) and (b) An initial 180° inversion pulse tips $M_0$ onto the $-Z$ axis. (c) During the delay ($\tau$) the magnetization recovers through spin–lattice relaxation. (d) A 90° excitation pulse is used to flip $M_z$ onto the $XY$ plane for detection.

## 3. Spin Echo

The $T_2$ measurement requires the ability to differentiate the true $T_2$ of the sample from the effects of $B_0$ inhomogeneity. The dephasing of the transverse magnetization caused by magnet inhomogeneity can be eliminated through the use of the spin–echo experiment. The positions of the magnetization vector during this experiment are diagrammed in Fig. 184. A 90° pulse applied along the $X'$ axis in the rotating frame of reference tilts $M_z$ from the $Z'$ axis onto the $Y'$ axis. Once placed in the transverse plane, nuclei residing in a region of higher $B_0$ field precess at a rate greater than that of the rotating frame of reference. Thus they move clockwise away from the $Y'$ axis. Likewise nuclei in a region of lower magnetic field move counterclockwise from the $Y'$ axis. Following a period of time $\tau$, variations in the strength of $B_0$ cause the individual nuclear spins to fan out in the $XY$ plane with a subsequent loss in the phase coherence of the transverse magnetization (Fig. 184). The application of a 180° pulse reflects the individual spins onto the $-Y'$ axis (Fig. 184). Now, however, the relative movement of the spins

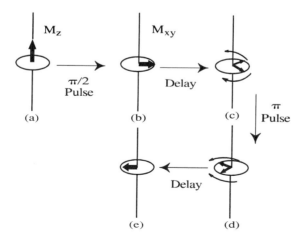

**Figure 184**  Spin–echo experiment: (a) Initial magnetization $M_z$ is tipped onto the $XY$ plane (b) using a 90° pulse. (c) In the $XY$ plane the individual spins fan out due to differences in local $B_0$ field strength. (d) A 180° RF refocusing pulse reverses the position of the individual spins in the rotating frame of reference. (d) Following the 180° pulse the individual spins continue to precess about the $Z$ axis according to the strength of the local magnetic field. (e) This causes the spins to refocus.

acts to refocus the transverse magnetization at a time $2\tau$. This refocused signal is termed a spin–echo. The spin–echo pulse sequence is represented as $90° - \tau - 180° - \tau - $ Acq.

For the refocusing to occur, nuclei must precess at the same rate and in the same direction following the 180° pulse as they did prior to the 180° pulse. Movement of the nuclei during the initial delay period may transport the spin to a region of differing $B_0$ field strength thereby changing the precessional frequency. Such movement may occur through Brownian motion or through coherent motions such as laminar flow. These displaced nuclei are not refocused by the 180° pulse and therefore do not contribute to the spin–echo signal. Signal loss due to diffusion can be minimized if $\tau$ is kept short. This is accomplished by employing a series of 180° refocusing pulses separated by short $\tau$ delays, which generates a series of echoes. The intensity of sequential echoes decreases exponentially as a result of spin–spin relaxation. This multi-echo experiment minimizes the effects of diffusion and $B_0$ inhomogeneity in the measurement of the $T_2$ relaxation time. Under some circumstances it is desirable to sensitize the spin echo to motional processes. In these cases magnetic field gradients are applied during the delay periods. The phase shift induced by movement through a magnetic field gradient can be used for the determination of diffusion coefficients, and for the imaging of flow as in angiography.

## K.  Overview of 2-D FT Experiments

The pulse sequences described thus far are designed to sensitize the NMR signal to a single variable. In the case of the progressive saturation and inversion recovery experiment, this variable is the $T_1$ relaxation time. For these one-dimensional studies the FID is collected as a function of time and converted to a frequency spectrum using a one-dimensional fast Fourier transform algorithm. Using multiple-pulse experiments, however, it is possible to determine signal intensity as a function of more than one variable. In multiple-pulse experiments the sequence can be divided into preparation, evolution, and detection stages. For example in the spin–echo sequence, the 90° pulse prepares the spin system by reflecting the magnetization onto the $XY$ plane. During the evolution phase the magnetization resides in the $XY$ plane, where it is acted on by $B_0$ inhomogeneity and spin–spin relaxation. The detection stage consists of acquisition of the spin echo. For certain experiments, the frequency and phase of the NMR signal depend on the nature of the preparation stage and the length of the evolution time. Although not sampled directly during the evolution time, information regarding the spin system can be encoded and observed in the phase of the FID. This allows two separate properties to be observed in a single NMR experiment. One feature is indirectly examined during the evolution time by encoding the phase of the signal, while the second is directly observed from the frequency of the detected signal.

In the two-dimensional NMR experiment a series a spin echoes are detected at incremental evolution times. Features that influence the magnetization during the evolution time induce a time dependent phase shift in the spin echo. A second Fourier transform converts this time domain signal into a second frequency for display. In the two-dimensional study the amplitude of the NMR signal is determined as a function of two frequencies. As we will see in a later section, current imaging sequences are based on two-dimensional Fourier transform techniques.

# II.   Generation and Detection of NMR Signal

## A.   Introduction

The simplest NMR spectroscopy or imaging system consists of a magnet with field gradients, a transmitter, a receiver, a coil/probe, and a computer that collects the data and orchestrates the timing and use of the entire system. The following sections explain the interaction of these components and discuss their functions and engineering requirements for high-quality clinical applications of NMR. A block diagram describing a basic imaging system is shown in Fig. 185.

## B.   The Magnet

The performance and quality of an NMR spectrometer or imager are dependent on the magnitude and homogeneity of the static or direct-current (dc) magnetic field. Initially most magnets used in the development of NMR were either permanent or electromagnetic (resistive). As higher magnetic field strengths were required, superconducting magnets appeared and are now the industry standard. Resistive magnets have become less important not only due to their lower field strengths but because they require excessive amounts of electrical current to maintain their fields. Low-field permanent magnets are attracting attention due to their low cost and progressively improving signal-to-noise ratio.

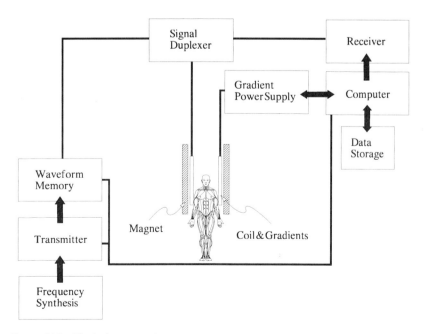

**Figure 185**   Block diagram of a magnetic resonance imager.

## 1. Superconducting Magnets

Superconductivity is the interesting property of certain metals that when cooled to liquid helium temperatures become perfect conductors. By a perfect conductor we mean that the electrical resistance in the wire is not just small, but is absolutely zero. Initially production of a high magnetic field with this technology was not only limited by the critical temperature of their superconductivity, but also by a unique critical magnetic field magnitude. These field magnitudes were very low. Fortunately a new class of alloys of the metal niobium was discovered to be able to carry very large current densities. This increase in current-carrying capacity resulted in today's high-field spectroscopy and imaging systems. Curiously it is the relation among critical temperature, field, and current density that is the obstacle to using recently discovered liquid-nitrogen-temperature superconductors for high-field magnets.

Since most metals exhibit near-zero resistance when they are cooled to temperatures close to absolute zero or 0 K, resistive coils can become superconducting when immersed in liquid helium, which has a boiling temperature of 4.2 K. No additional power is needed to maintain the magnetic field once the field is established. The niobium–titanium alloys used for fabricating coils in superconducting magnets have a critical temperature between 10 and 15 K, that is, under this temperature the material is superconducting, having minimal resistance. The current density $(A/m^2)$ in the conductor, magnetic field produced by the current, and the critical temperature are all interdependent. At constant current density, the magnetic field increases as temperature decreases. At constant temperature, the sustained current density decreases as magnetic field increases. Therefore, the coil may lose its superconductivity whenever any one of these parameters exceeds a threshold.

In a superconducting magnet, the coil is mounted in grooves cut on an aluminum cylinder with more conductors near the ends to produce a more uniform field. The coil and cylinder are suspended in liquid helium. The helium container is suspended in a vacuum, which in turn is enclosed by liquid nitrogen with a boiling temperature of 77 K. This structure is surrounded by another layer of vacuum. A picture and the schematic diagram of a superconducting magnet are shown respectively in Fig. 186 (a) and (b). The heat shields are specialized structures designed to prevent heat loss due to convection and radiation. Under reasonable conditions, once the field is established, the current density can be maintained over a long period of time. It would take more than a thousand years for a helium-cooled magnet to lose one half of its field strength. The field drops due to small nonzero resistance of the junctions between individual pieces of superconductors. The wire must be spliced in sections since it is impossible to draw a single conductor that is of sufficient length.

Superconducting magnet windings are not simply wound from single strands of niobium alloy wire. First, at temperatures above their critical temperature, niobium alloys have high electrical resistance, making them unsuitable for carrying high current loads unless near absolute zero. Second, these alloys are very brittle and difficult to handle. These problems have been largely overcome by encasing many filaments of the niobium alloy into a matrix of very pure copper metal. This is performed by drilling holes into a thick cylinder of copper into which rods of niobium alloy are placed (Fig. 187). The entire cylinder or "billet" is drawn into wire. The process can be repeated to increase the number of filaments in the wire. In a new 3.0-tesla whole-body magnet the winding is made of an 0.8-mm

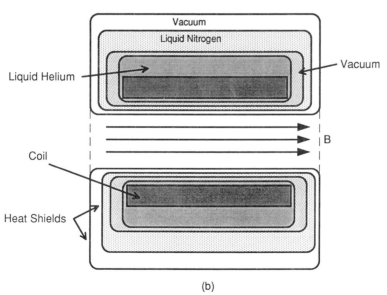

(b)

**Figure 186** (a) Superconducting magnet (1.0-meter bore) showing its external cryostat (Courtesy of IGC, Guilderland, New York). (b) Drawing of the cross section of a similar magnet showing the positions of the winding, cryogens, and dewars.

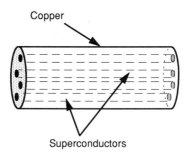

Copper

Superconductors

**Figure 187** Cross section of a "billet" or copper cylinder containing rods of superconducting alloy. The billet will be extruded into a multiple-core wire used to wind superconducting magnets.

diameter copper wire containing 52 individual filaments of niobium alloy. This wire provides both structural support and a low resistance path for current density when the temperature of the wire is above superconducting critical temperature.

Additional difficulties have been presented by the sheer size of the magnets in whole-body NMR imaging and spectroscopy. The magnetic field exerts a force ($F$) per unit volume of current-carrying conductor. State-of-the-art whole-body magnets at 3 to 4 tesla field strength may contain well over a ton of current-carrying conductor. Faraday's law suggests that the magnetic field could be pictured as a collection of lines of force that are repelling each other. The magnetic field orthogonal to the lines of force exerts a pressure $B^2/2\mu_0$ on the windings of the magnet where $\mu_0$ is the permeability of free space. At 4 tesla (T) this pressure is equivalent to 62 atm (atmospheric pressure). This value is within an order of magnitude of the yield strength for many soft metals. Without the proper engineering these forces can break the cryostat, deform the magnetic winding, degrade magnet performance, or mechanically fail without warning, thus compromising the safety of patients or the system operator.

Superconducting magnets are capable of producing large, stable magnetic fields of greater than 10 T with excellent uniformity in a volume 2 cm in diameter for spectroscopy studies. Superconducting magnets as large as 4 T, with bore diameters of 1 m have been constructed for imaging scanners. Once the current in the superconductor has been established, no external power is needed to maintain the magnetic field. However, if the system is disturbed, a sudden loss of superconductivity or quenching can occur. A number of mechanisms can result in quenching. In wires where there is a transverse magnetic field, surface currents may be induced to shield the conductor from or to counterbalance the magnetic field. These surface currents produce heat. If the surface currents are not strong enough to completely counterbalance the magnetic field, additional currents are generated in the interior of the conductor, which also produce heat. The heat is conducted to the surface of the conductor. If the heat cannot be rapidly dissipated, temperature will rise, reducing the current density that the conductor can support. A vicious cycle results. This problem can be partially solved as previously discussed by embedding many filaments of alloy conductor in copper, as shown in Fig. 187. Another potential cause of quenching is that the electromagnetic interaction among conductors can result in attractive forces among the conductors, as previously discussed. Slight movements of the conductors due to these forces produce friction and heat.

As imaging techniques have become more sophisticated, higher requirements are needed for magnet homogeneity. Homogeneity is specified by the change in magnetic field in parts per million (ppm) over the diameter of spherical volume (dsv). Typical imaging magnets of one-meter bore size specify homogeneity of ±5.0 ppm at 50 cm dsv. This means that the field does not increase or decrease over 5.0 ppm within a sphere centered at the magnet isocenter having a diameter of 50 cm. Significant magnet inhomogeneity can cause shape distortion and nonuniform image intensity. Difficulties may be experienced in imaging regions at some distance from the isocenter. This would include imaging of the shoulder and images requiring three-dimensional imaging techniques. In smaller volumes near the isocenter the field homogeneity is dramatically improved. Specifications at 10 cm dsv are often as good as ±0.05 ppm.

The magnetic field produced by superconducting magnets can be very strong. As a result, the fringe field surrounding the magnet sometimes can be significant although the field strength drops off rapidly as a function of distance from the magnet. Nevertheless, great care must be taken in ascertaining that no ferromagnetic objects are brought to the vicinity of the magnet. A so-called "missile effect" can occur when these objects are pulled to the magnet at high speed by the strong magnetic force. Federal regulation requires there should be controlled access of the public in areas where the magnetic field is higher than 5 gauss. A typical superconducting MRI magnet situated in a shielded room is shown in Fig. 188. A sagittal view of the brain is shown on the console display. The patient is moved in and out the gantry by a motorized bed. Since the opening of the gantry is long and deep and the noise produced by the magnet can be very bothersome, earphones may be provided to the patient to ease the tension.

In the past few years, great strides have been made in developing new superconducting materials that retain superconducting property even at temperatures higher than the boiling temperature of liquid nitrogen (Hatfield and Miller, 1988). A current problem of this technology is that the current density that can be produced by these materials at these temperatures is still too small to be of practical use. If this problem can be circumvented, liquid helium, which is relatively rare and expensive, can be replaced by liquid nitrogen. This would mean considerable savings in the operating cost of a superconducting magnet.

## 2.  Permanent Magnets

Less than 3% of the more than 2000 magnets operating in the United States are low-field systems. These systems operate at fields far below superconducting magnets having field strengths of 0.064–0.3 tesla. Uniform magnetic fields can be achieved at the center of a permanent or electromagnet, as shown in Fig. 189 if the width of the gap is much smaller than the lateral dimensions of the magnet.

The advantages of low-field MRI systems are economic. First, the initial cost of purchase of the system is less than one-half of a corresponding high-field system. Second, they are less expensive to install. The magnet's stray field is significantly reduced, so they take up less space. They often weigh only 6 tons compared to the 20–40 tons of a high-field system and its associated shielding. Finally, they are cheaper to operate since no cryogens are required for their maintenance.

One important noneconomic advantage of low-field magnets is that they are more open around the patient (Fig. 190). It is estimated that approximately 5% of patients refuse to be placed in an MRI magnet due to claustrophobia. Open-space,

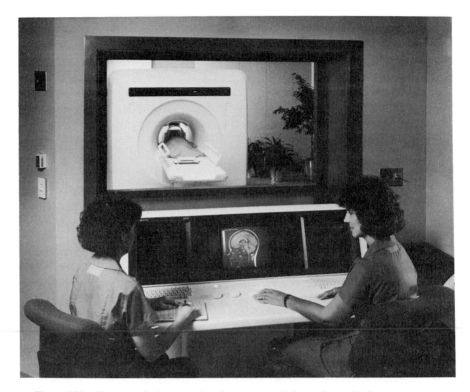

**Figure 188** Photograph showing the observation window of a radio-frequency shielded room (Courtesy of GE Medical Systems, Milwaukee, Wisconsin). Electromagnetic radiation of 0–100 MHz is reduced by 100 dB inside the magnet area.

low-field permanent magnets are in almost all cases acceptable or tolerable to this patient group.

Low-field systems suffer, however, from lower signal sensitivity. This means the image quality is not as good. In recent years, substantial improvements have been made in image quality. For some diagnoses, low-field MRI is more than adequate particularly in shoulders, knees, and other skeletal muscular applications.

## C. Room Temperature Magnetic-Field Gradients

Room temperature magnetic-field gradients are used in two ways in NMR imaging. They are used either to compensate for magnetic-field inhomogeneities of the magnet or to provide an encoding magnetic-field gradient to make an image. In either case these gradients are arrangements of loops of copper wire designed to produce well-defined gradient patterns in the bore of the magnet.

Rarely is a magnetic field perfectly homogeneous and would not remain so after the magnetic susceptibility of the human body is placed within the bore. Magnetic gradients or shim coils are used to optimize the magnetic homogeneity. These adjustments include linear gradients in all three axes as well as higher-order gradients whose behavior may vary up to the fourth power of the gradient.

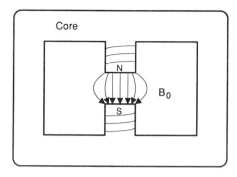

**Figure 189** Diagram showing the magnetic field produced from a permanent or iron-yoked electromagnet.

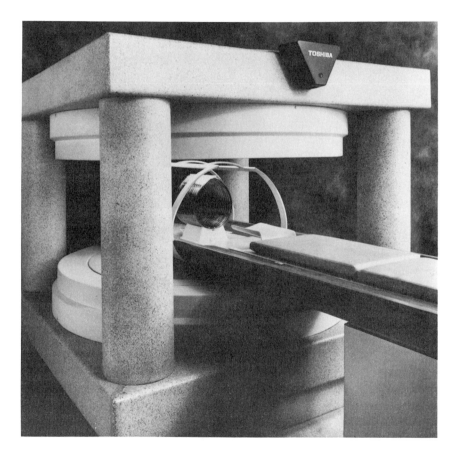

**Figure 190** Photograph of the orientation and patient access of a permanent magnet used for MR imaging (Courtesy of Toshiba American Medical Systems, Tustin, California).

Using this method most inhomogeneities within the central field in the bore of the magnet may be corrected. These compensations are generally not larger than $\pm 10$ ppm.

Additional linear gradient sets located along the three central axes of the magnet are used to encode the space inside the bore to produce an NMR image. These are the gradients used for slice selection, frequency encoding, and phase encoding of the NMR signal. Typically these gradients are a little stronger than shimming gradients, producing linear fields of about 10 mT/m. More important, these gradients must respond quickly to the command of the computer to achieve their maximum strength and quickly return to a zero value. The ability to "turn on" and equilibrate quickly is described by the rise time to achieve the required field strength. For clinical NMR systems these values are 500 to 1000 $\mu$sec. Unfortunately the gradient loses energy into surrounding metal structures and often undershoots their designated values by a few microseconds. This can be controlled to some extent by reducing the coupling between the gradient coils and the magnet bore by decreasing the coil diameter and physically distancing the coil from metal near the magnet bore.

Difficulties also arise when the pulsed gradients are turned off. During the gradient pulse, the linear magnetic-field gradient couples electrically to any metal surface. This surface is usually the bore tube and cryostat of the magnet. Eddy currents are rapidly formed on the metal's surface. When the gradient is turned off the magnetic field does not decrease sharply but decays slowly due to the eddy currents and coupling. Often a small fraction of the voltage will oscillate positively and negatively for many milliseconds, resulting in a degradation of image quality. Further, the time needed to equilibrate the gradients proves to be the limiting factor for the system's shortest achievable echo time (TE). Gradient stabilization times establish limits of speed for most fast imaging techniques.

Capacitive/resistive time-constant circuits can be designed to control the electrical decay of this rapidly rising voltage. Several of these circuits having different time constants are used to shape the decay voltage into a normal square wave. This adjustment is called the preemphasis of the pulsed gradient.

Eddy current formation also can be controlled by actively shielded coils. In this scheme extra gradient windings are arranged on the outer diameter of the gradient coil sets. The current in these active shielding coils circulates in the reverse direction to the gradient coils to provide an opposing magnetic field. The geometry of these coils is arranged to decrease the field at metal surfaces near the bore tube to zero while leaving the gradient field inside the bore unchanged. If there is no pulsed magnetic-field energy to couple to the bore tube or the walls of the cryostat, rise times and eddy currents are substantially reduced. The disadvantage of this technique is that it requires additional space for the windings, which decreases the open diameter cross section used to accommodate the patient. Compromises between gradient performance and usable space inevitably occur.

## D.   The NMR Coil/Probe

### 1.   Homogeneous Field Coils

Perturbation of the magnetization vector is accomplished with an external RF energy source at the same frequency as the Larmor frequency of the nuclei. The RF energy is coupled to the object containing these nuclei by an RF inductor (NMR coil). Since NMR devices are typically operated in pulsed mode, the same coil may be used following excitation for detecting the NMR signal emitted by the nuclei returning to the lower energy state.

The coil design with NMR devices is different from an RF antenna, where a large fraction of the emitted energy is irradiated into the far field of the radiator. The RF coil in NMR devices needs to retain the energy in the near field with little radiation into the far field. Some of the major requirements of an RF coil in an NMR imaging system are that it must be large enough to accommodate body parts, be able to produce a uniform RF field, have minimal loss, and not interfere with other parts of the system. A number of RF coil designs including the solenoid, the saddle coil, and the "birdcage" coil have been used in NMR devices (Partain *et al.*, 1988). In general no matter what coil design is used, high spatial uniformity and signal-to-noise ratio cannot be achieved simultaneously. An increase in spatial uniformity results in a decrease in signal-to-noise ratio and vice versa. This is exemplified by the utilization of surface coils, which increases the signal-to-noise ratio from a limited sample volume at the expense of spatial uniformity over an extended volume.

Multiple-turn solenoids can produce a uniform coil field if the length of the coil is much greater than its radius. They are not desirable for high-frequency operation because of their high inductance and stray capacitance. In addition the orientation of the $B_1$ field requires the solenoid to be placed perpendicular to the $B_0$ axis. This makes entry difficult in resistive and superconducting magnets.

The Helmholtz coil is often used in whole-body applications because it provides good radio-frequency homogeneity while providing open access along the bore of the magnet. The Helmholtz coil's field is homogeneous because it crudely approximates a spherical coil as illustrated in Fig. 191(a) (Everett and Osemikhian, 1963). Ideally an infinite number of coil pairs arranged on a sphere would produce a perfectly homogeneous magnetic field. The Helmholtz coils have been frequently used for generating a low dc magnetic field inside the sphere. For transmitting and receiving RF fields, the resulting magnetic field produced in the coil's $X$ direction is most uniform if the current density distribution in the wire windings in the $XY$ plane is sinusoidally weighted, as shown in Fig. 191(b). Figure 191(c) shows a practical realization of this where two windings are used to approximate a sphere. When the return paths are provided for the current, the coil looks like a saddle and is therefore called "saddle coil" (Partain *et al.*, 1988). The dotted and hatched circles indicate respectively current going out and into the $XY$ plane.

The same RF coil may be used to transmit and to receive the RF signal, but sometimes it would be desirable to have a separate coil to receive. Using separate coils allows a geometric configuration of individual coils be optimized for transmission or reception. When coils of smaller cross section are used to receive signal, they couple more strongly to the sample yielding more signal intensity.

## 2. Surface Coils

Surface coils are simple wire loops of various sizes tuned and matched to the frequency of interest. These coils overcome two technical difficulties in NMR spectroscopy and imaging. First, surface coils are useful because they localize signal reception to organs or structures of interest. Second, they provide greater sensitivity since their radio-frequency magnetic field is more intense than larger homo-

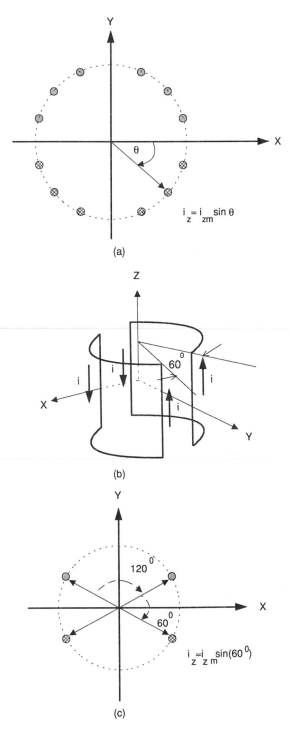

**Figure 191** Development of the Helmholtz coil. (a) Sinusoidal arrangement of wires with positive and negative current that produces a homogeneous magnetic field. (b) Simplification to four current-carrying wires with a circular current return path. (c) Simplified cross section of the four-wire distribution showing their optimal angular orientations.

geneous coils. Unfortunately, the magnetic field produced by surface coils decreases rapidly with distance and thus does not penetrate deeply into the tissue. The axial radio-frequency magnetic field ($B_{1a}$) decreases as

$$B_{1a} = \mu_0 I r^2 / (z^2 + r^2)^{1.5}$$

where $\mu_0$ is the permeability of free space, $I$ is the current in the wire loop, $r$ is the loop's radius, and $z$ is the axial position at which the field is measured. At an axial distance of about one coil radius, the surface coil's magnetic field is only 35% of its field intensity at the coil center.

Typically in standard clinical imaging, surface coils are used in the "receiver only" mode. In this procedure the homogeneous radio-frequency field of the large-body coil is used to stimulate the entire sample. However, when used in this way the surface coil receiver will produce a 30% increase in signal-to-noise ratio and slightly better acquisition of signal from regions deeper into the sample. Surface coils are most often used on the knee, shoulder, neck, spine, and eye (orbit). They are also widely employed in *in vivo* NMR spectroscopy.

## E. The Transmitter

All pulse NMR applications demand a short but intense high-power RF magnetic field in order to perturb the alignment of the nuclear magnetization to obtain a signal. A frequency generator provides a source of continuous RF input. This signal is gated to allow the computer to control the timing of the pulse at the proper times. The gated signal is fed to a power amplifier, which may boost the signal to 5–20 kW of output power. This amount of power could be very damaging if the power remained switched on for any reason. Therefore, most conventional imagers have several systems (power meters, directional couplers, RF fuses) to sense continuous irradiation and provide an automatic shutdown of the amplifier. The effects of RF heating are discussed in Section VI.

The duration of the RF pulse is critical. Errors in the value of the pulse time can lead to significant loss in the signal-to-noise ratio. This is because the signal amplitude is proportional to the sine of the tilt of the nuclear magnetization. Inaccuracies in the 180° flip angle used to generate spin echoes (discussed in the previous section) will also result in large signal losses. Finally, the inverse of the pulse duration defines the bandwidth of the RF irradiation.

In current MRI systems the computer is able to control both the amplitude and phase during the several milliseconds or less that the transmitter is on. The actual hardware that performs this function in the computer is called waveform memory. Complex manipulation of the transmitter pulse's amplitude and phase can result in an improvement of the frequency profile of the excitation pulse. This technique can correct for $B_1$ inhomogeneity inherent to the coil, improve bandwidth stability, or define specific frequency performance for specialized pulse sequences.

## F. The Receiver

The receiver consists of a number of individual components. The incoming NMR signal has already been amplified by the preamplifier as it emerges from the coil. In the receiver the signal is first demodulated. This means that the carrier frequency, which is either near or equal to the resonance frequency, is removed. The resulting low-frequency signal is filtered to remove unwanted noise outside the chosen bandwidth. This improves the final signal-to-noise ratio. Next, the signal is split and passed through a pair of phase-sensitive detectors. The output of these detectors is sensitive to the difference between the measured NMR signal and the reference signal of the spectrometer. Two of these detectors are used, but in one detector the reference signal is phase-shifted by 90°. This is known as quadrature detection. In this manner two measurements are made of the decaying NMR signal, but are now measuring the noise component for both the positive and negative portion of the bandwidth. This method of sampling improves the signal-to-noise ratio by a factor of $\sqrt{2}$.

## G. Data Acquisition

The two analog signals are filtered once more to eliminate aliasing and are digitally sampled. The data is sampled (on the order of 50 kHz) so that there are at least two data points for the highest frequency cycle in the sample. This is referred to as the Nyquist condition. If this condition is ignored frequencies may appear in the Fourier transformed data set at one sampling rate below the true frequencies. These aliased signals can be confusing in NMR spectroscopy and can create severe signal-folding artifacts in images. The final signal is stored in an analog-to-digital converter that generally contains 16-bit word depth. This means that the maximum signal intensity can be stored at a value of $2^{16}$ or 65536. The numbers from these registers record the free induction decay (FID) and are then stored on a computer's hard disk.

# III. Imaging Methods

## A. Introduction

In the NMR imaging experiment spatial information must be converted to frequency information. For a three-dimensional object, each point in space must occupy a unique frequency. Through the Zeeman effect linear magnetic-field gradients spatially encode the resonant frequency of the nuclei in a linear fashion. This necessitates the use of three orthogonal gradients. These gradients are defined as the slice selection, phase encoding, and frequency encoding gradients. As we will see in the following sections these gradients are applied during the preparation stage, the evolution stage, and the detection stage of the pulse sequence, respectively.

### 1. Slice Selection

Nuclei residing in a two-dimensional slice can be selectively excited from a three-dimensional sample using the combination of a slice selection gradient and a selective RF pulse. Although slice selection may occur during any period in which an RF pulse is applied, it is common to select the slice during the preparation phase of the imaging sequence. As diagrammed in Fig. 192, slice selection requires the simultaneous application of a linear magnetic-field gradient and selective RF energy. In this example a cylindrical tube of water is placed parallel to the

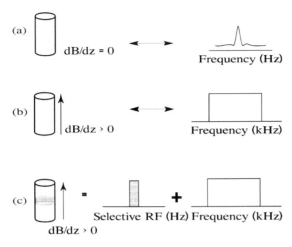

**Figure 192**  Slice selection: (a) A cylinder of water placed in a uniform magnetic field. (b) A cylinder placed in a linear magnetic-field gradient. (c) When a cylinder of water is placed in a linear magnetic-field gradient and irradiated with a selective RF pulse, only those nuclei residing in the selected slice are excited.

Z axis in a uniform magnetic field. Without an external magnetic-field gradient all of the protons in the water sample would have the same resonant frequency. As illustrated in Fig. 192(a), an NMR spectrum taken from this sample would consist of a single narrow line. The application of a linear magnetic-field gradient along the Z axis, as shown in Fig. 192(b), causes a linear dispersion of the resonant frequencies. As a result of the applied gradient, the resonant frequency of an individual proton is directly proportional to its position on the Z axis. An NMR spectrum of this sample would ideally demonstrate a rectangular frequency profile in which the NMR signal is dispersed over a frequency range of several kilohertz. The actual range of frequencies covered by the sample is proportional to the strength of the applied field gradient. If a selective 90° RF pulse is applied in the presence of the magnetic-field gradient, a cross-sectional slice of nuclei perpendicular to the Z axis is excited, as in Fig 192(c). Recall that a selective RF pulse has a narrow frequency bandwidth that typically covers several hertz. The position of the slice can be varied by changing the carrier frequency of the RF pulse. The slice thickness is dependent on the magnitude of the applied magnetic-field gradient (dB/dz), and the frequency bandwidth of the selective RF pulse. For a constant gradient strength, the thickness of the excited slice is varied by changing the length of the RF pulse (PW), which is inversely proportional to the bandwidth of the selective pulse.

The timing of the gradient and RF pulses, or pulse sequence, for slice-selection is illustrated in Fig. 193. Often it is desirable to observe more than one slice in a sample. In these situations a multislice sequence is employed, in which multiple slices are selected and observed. In these cases it is critical that the excitation and refocusing RF pulses act on the same slice of the sample. This requires slice selection during both the 90° excitation pulse and the 180° refocusing pulse of a spin–echo experiment and places additional requirements on the slice selection component of the experiment. Nuclei residing in the transverse plane will undergo a phase shift when exposed to a change in magnetic field. As we will see later in

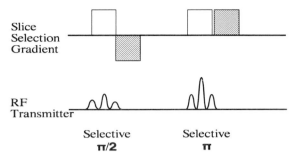

**Figure 193** Slice selection sequence for selective 90° and 180° pulses. The shaded gradients are applied to correct the phase effects produced by the slice selection gradients.

this section it is critical that all phase encoding of the NMR signal be generated by the phase-encoding gradient. Additional gradients are needed to compensate for the phase effects of the slice selection gradients. These are shown as grey regions in Fig. 193. A finite amount of time is required for exciting the slice with a 90° pulse. The nuclei, which have been tipped into the XY plane, experience a phase shift from the slice-selection gradient. This phase shift can be corrected by reversing the slice-selection gradient during the initial delay period. By reversing the direction of precession relative to the rotating frame of reference, the spins are refocused in the XY plane. Likewise, time is needed to establish the slice-selective gradient prior to application of the 180° refocusing pulse. Since the 180° pulse acts to invert the direction of precession, this phase shift is corrected by using a positive gradient after the 180° pulse. This method of correcting phase distortions in the slice-selection is just one example. Many schemes for correcting phase distortions have been reported. In addition to refocusing through gradients it is theoretically possible to correct phase errors using self-refocusing RF pulses.

The ideal selective RF pulse would be shaped like a top hat in the frequency domain. For such a pulse, nuclei residing within the frequency bandwidth of the pulse would be irradiated with exactly the same amount of RF energy, and would undergo the same flip angle. Likewise, nuclei outside of the frequency range of the selective pulse would absorb no RF energy and would remain unperturbed by the selective excitation. Despite the continued search for such a pulse, it does not yet exist. In reality selective pulses contain a fringe region in which neighboring nuclei experience intermediate flip angles. Nuclei located in these regions or transition zones can generate spurious signals leading to artifacts in the image. Presently selective-pulse programmers use digital waveform memory. A digital time-domain signal modulates the amplitude of the RF pulse. The frequency profile of the resultant selective pulse is described by a Fourier transform of this time-domain signal. Commonly used time-domain signals employ Gaussian or sinc functions to approximate a top-hat frequency profile.

## 2. Frequency Encoding

Once a slice has been selected, the excited nuclei must be spatially encoded in the remaining two dimensions. In one dimension the frequency of the NMR signal can be spatially encoded by applying a linear field gradient during the detection stage, as illustrated in Fig 194. This gradient directly encodes spatial information

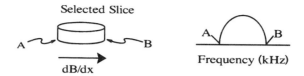

Selected Slice

Frequency (kHz)

**Figure 194** Frequency encoding of linear magnetic-field gradient. A and B correspond to positions in the selected slice. Their corresponding resonant frequencies are shown at the right.

along the applied axis. In the example shown, the gradient (dB/dx) is applied along the $X$ axis.

The finite period of time required to establish the frequency-encoding gradient induces a phase shift in the NMR signal. As indicated in Fig. 195, this phase shift can be compensated by applying a positive gradient prior to the 180° refocusing pulse. As reviewed in the discussion on the spin–echo experiment, diffusion or movement of nuclei over the time period between the applications of the gradient pulses decreases the refocusing of the 180° pulse, resulting in a loss of signal. An alternative method for correcting phase artifacts is to apply a negative gradient just prior to the positive frequency-encoding gradient. The refocusing of the nuclear spin produced by gradient reversal also produces a signal termed a gradient echo. A gradient echo is also produced in the case where the gradient pulse is applied prior to the 180° pulse. To minimize destructive interference from these signals, the timing parameters and gradient strengths are adjusted to produce complete overlap of the spin echo, caused by the 180° pulse, and the gradient echo, caused by the pair of gradient pulses.

## 3. Phase Encoding

Spatial encoding of the second dimension in the selected slice occurs through phase-modulation of the NMR signal. A phase-encoding magnetic-field gradient is applied during the evolution stage of the sequence in which the nuclear spins are precisely in the $XY$ plane. Although phase encoding could be achieved by maintaining a constant gradient strength and incrementing the evolution time for sequential scans, each acquisition would have a different time between excitation and detection of the echo. For such a sequence the $T_2$ dependency of the signal

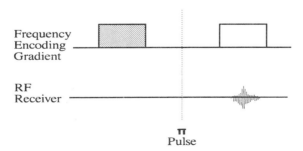

Frequency Encoding Gradient

RF Receiver

π Pulse

**Figure 195** Frequency encoding of a pulse sequence. The frequency-encoding gradient is applied during acquisition of the NMR signal. The shaded field gradient is applied to correct the phase effects generated by the frequency-encoding gradient.

would be nonuniform along the phase-encoding axis. Instead the time between the 90° pulse and detection, called the time of evolution or TE, is kept constant and the strength of the phase-encoding gradient is changed in incremental steps. The phase-encoding pulse sequence is diagrammed in Fig. 196, where the horizontal lines indicate that the amplitude of the gradient pulse is increased with each sequential scan. The incremental change in the gradient strength induces a periodic phase modulation of the NMR signal (Fig. 197). The frequency of this modulation is directly proportional to position along the phase-encoding axis, and can be extracted from the data using a second Fourier transform.

Unlike frequency encoding, which occurs with the application of a single gradient, phase encoding requires the combined effect of numerous scans. An image having 256 pixels in the phase-encoding axis requires 256 independent phase-encoding steps. The process of phase encoding is responsible for the long time periods required to generate an image when standard 2-D FT sequences are used. These phase-encoding steps may be acquired in any order; however, it is common to increment the gradient strength starting at the most negative value. Since the phase information is symmetric about the zero point it is possible to collect only half of the phase-encoding steps and then calculate the remaining half prior to transforming the data. This process known as half-Fourier acquisition decreases the number of scans, and therefore the imaging time, by one-half. However, since fewer scans are averaged, there is a subsequent decrease in the signal-to-noise ratio of the image.

## B.  Spin–Echo Imaging

The spin–echo imaging sequence is the imaging correlate of the standard spin–echo spectroscopy experiment described earlier. It forms the basis for many of the standard sequences used in clinical MRI. The complete spin–echo imaging sequence, diagrammed in Fig. 198, consists of the combined pulse sequences of slice selection, frequency encoding, and phase encoding, illustrated in Figs 193, 195, and 196. The initial selective 90° pulse excites the protons residing in the sample. Within this selected plane the magnetization vector associated with these protons is rotated from its equilibrium position on the $Z$ axis into the $XY$ plane. Once in the $XY$ plane magnetic field gradients can act on the nuclei to change the precessional frequency, and thus encode the phase of the signal. This phase information persists in the signal after the pulse gradient has been terminated. A 180° RF pulse refocuses the dephasing that results from magnetic field inhomogeneity. Since the phase-encoding gradient is applied only in the period prior to the 180°

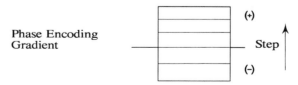

**Figure 196**   Phase encoding of a pulse sequence. The phase-encoding field gradient is applied while the magnetization is precessing in the $XY$ plane. The horizontal lines indicate that the gradient strength is to be incremented for sequential scans.

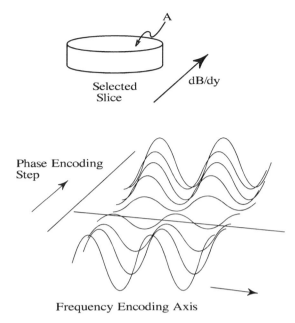

**Figure 197**  Effect of the phase gradient for a proton located at position A in the selected slice of the sample. Since the frequency-encoding gradient is constant for all scans, the resonant frequency of this proton is the same for all scans. The incremental change in the magnitude of the phase-encoding gradient changes the phase of the signal for sequential scans. This phase change generates a periodic modulation of the signal in the second dimension. The frequency of this modulation is a function of position along the phase-encoding gradient (*dB/dy*).

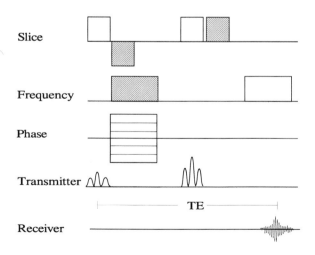

**Figure 198**  Spin–echo imaging pulse sequence. The period TE or time to echo is the time during which the magnetization resides in the *XY* plane. The time to repetition or TR is the time between sequential 90° RF excitation pulses.

pulse, the refocusing pulse does not compensate for the phase shifts induced by this gradient. The spin–echo signal is frequency encoded by application of a magnetic-field gradient during acquisition. This sequence of events is repeated $N$ times for $N$ phase-encoding steps in which the magnitude of the phase-encoding gradient is incremented following each acquisition. After all of the echoes have been collected, a two-dimensional Fourier transform is performed to transform the time domain signals into a two-dimensional frequency map. The NMR image is thus a spatial representation of the NMR signal amplitude as a function of two frequencies.

Two timing parameters refer to specific periods between RF pulses. These are the time to repetition (TR), and the time to echo(TE). The magnitude of these values are varied to achieve contrast in the NMR image. The TR interval is the total time between sequential 90° excitation pulses for a specific slice. It is analogous to the delay ($\tau$) in the progressive saturation experiment. During this interval the spin system is recovering magnetization along the $Z$ axis via spin–lattice relaxation. The TR interval can be adjusted to sensitize the NMR signal to the $T_1$ time of the sample. Such sequences are said to be "$T_1$-weighted." In Fig. 180 is a plot of $M_z$ versus $\tau$ for two tissues with differing $T_1$ times. The nuclei in compartment A recover their $M_z$ rapidly, characterized by a $T_1$ of 1.0 sec, compared with the long $T_1$ of sample B, which is 3.0 sec. If this sample had been imaged using a sequence with a long TR of 10 sec, the nuclei from both compartments would have returned to their equilibrium conditions. The signal intensity, and therefore the contrast between compartments, would be primarily determined by the proton density of the two compartments. However, if the TR is reduced to 1 sec, the nuclei in compartment A recover a significant portion of $M_z$ while the nuclei in compartment B are sampled under conditions of partial saturation. In the subsequent image compartment A appears bright due to the $T_1$-weighting of the sequence. In spin–echo imaging $T_1$-weighting may be added to a sequence by reducing the TR interval. This is analogous to decreasing the $\tau$ delay in the progressive saturation spectroscopy experiment. On these images, regions with short $T_1$ appear brighter, while those with long $T_1$ are darker.

Significant $T_1$-weighting can be added to an imaging sequence if the 90° excitation pulse is preceded by a 180° inversion pulse. This is analogous to the inversion recovery experiment described for the measurement of $T_1$ values. The time between the 180° inversion pulse and the 90° excitation pulse is called the time of inversion or TI interval, which is analogous to the $\tau$ delay in the inversion recovery spectroscopy experiment. During the inversion interval the magnetization vector returns from its maximal negative value to its original position along the positive $Z$ axis via spin–lattice relaxation. The effect that this process has on the observed NMR signal is a complex function of the $T_1$ of the sample and the value for TI. Recall that the inversion recovery spectroscopy experiment uses phase-sensitive detection. In this mode the observed absorption peak changes from negative to positive for increasing TI values. In conventional two-dimensional NMR experiments, magnitude-calculated spectra and images are obtained to compensate for the complex phase distortions present in the signal. Since the phase information is not utilized, only the magnitude, not the direction of the $M_z$ vector, is sensed in the inversion recovery image. Therefore as the magnetization vector passes from negative to zero, the image intensity actually decreases. It then increases as it passes from zero to a positive value. Note that a value of TI that coincides with the time at which the magnetization vector passes through zero or null point may be chosen. In this case no signal would be detected. This has useful applications such as suppressing the signal from fat, which can produce artifacts in the image. In addition this process may be used to provide significant $T_1$-weighting. A major disadvantage of the inversion recovery sequence is the long delay between acquisitions to allow the magnetization vector to fully relax. Because of this delay, inversion recovery images require a long time period compared to $T_1$-weighted images acquired with progressive saturation.

During the TE period of the experiment, the magnetization vector resides in the *XY* plane. While in the transverse plane, the magnitude of this vector decreases due to spin–spin relaxation. Therefore the TE period can be varied to sensitize the signal amplitude to the $T_2$ of the sample. Sequences with long TE times are said to be $T_2$-weighted, since the signal amplitude is sensitive to the $T_2$ of the sample. As demonstrated in Fig. 181, samples with short $T_2$ will lose their transverse magnetization and therefore appear dark on sequences with long TE periods. Because a finite period of time is required to spatially encode the transverse magnetization and refocus it with a 180° pulse, all images have some degree of $T_2$-weighting. For the image intensity to be sensitive to differences in the $T_2$ of the tissue, the TR must be longer than twice the longest $T_1$. This minimizes the $T_1$ dependence of the signal, which would dominate under conditions where TR is less than or equal to the $T_1$ of the sample. In addition images with $T_2$ sensitivity may be obtained using a multi-echo sequence. In this sequence short TE values are used to minimize signal loss from diffusion and motion. Since the primary mode of signal loss is through spin–spin relaxation, the signal intensity of the later echoes is very dependent on the $T_2$ of the sample.

In the standard spin–echo imaging experiment signal intensity, and therefore contrast between tissues, can be made sensitive to spin density, $T_1$, and $T_2$ properties of the sample. As demonstrated by the spin–echo MR images shown in Fig. 199, the sensitivity of signal intensity to these parameters generates excellent soft tissue contrast. As summarized in Table XVI, this sensitivity is optimized by adjusting the TR, TE, and when using inversion recovery, TI, intervals. The interdependence of these parameters means the MR image contrast is very sensitive to acquisition parameters. While this adds versatility to the modality, it also complicates image interpretation and sequence development.

## C.  Gradient–Echo Imaging

In addition to refocusing the transverse magnetization with RF energy, nuclear spins may be rephased using bipolar magnetic field gradients. This forms the basis for the gradient–echo imaging sequence diagrammed in Fig. 200. Following the initial RF excitation pulse a negative-read gradient is applied that dephases the transverse magnetization. The direction of the read gradient is then reversed during signal acquisition to frequency encode the NMR signal. In the rotating frame of reference, the reversal of the read gradient causes the spins to move in the opposite direction and therefore refocus. The duration and magnitude of the bipolar-read gradient are adjusted to refocus the spins at the center of the acquisition period. The refocused signal is called a gradient echo to differentiate it from spin echo, which is generated by a 180° refocusing pulse. The slice-selection and phase-encoding components of the gradient–echo sequence are identical to the spin–echo experiment described earlier. The major difference in the gradient–echo pulse sequence is the absence of the 180° RF pulse, which results in less satura-

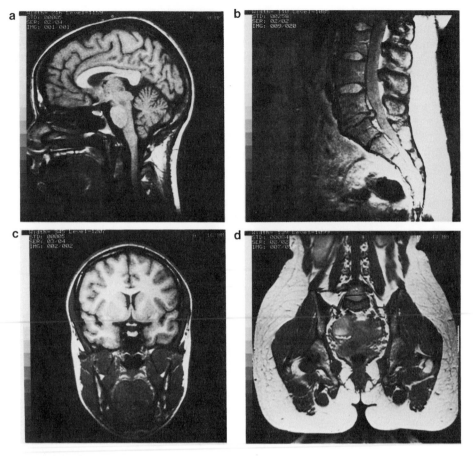

**Figure 199**  (a) Image of a head demonstrating excellent soft-tissue contrast and fine anatomic detail obtainable at high magnetic field strengths using 3 mm slice thickness.
(b) Image of lumbar spine in the sagittal plane showing a herniated disc at the L5-S1 level.
(c) A 3-mm image of the head in the coronal plane showing the pituitary fossa. The ability to achieve thin slices and to vary the image plane allows visualization of the pituitary stalk and small tumors of the pituitary gland.
(d) Image of the abdomen showing the psoas muscle, uterus, bladder, and pelvic bone marrow. The bright area corresponds to the region occupied by a pelvic cyst. All images were acquired on a 1.5-tesla Signa system at General Electric's MR Development Center in Milwaukee, Wisconsin.

## Table XVI
Dominant parameters that affect image contrast with spin–echo imaging sequences

|            | Short TE       | Long TE     |
|------------|----------------|-------------|
| Short TR:  | $T_1$          | $T_1$, $T_2$ |
| Long TR:   | Proton density | $T_2$       |

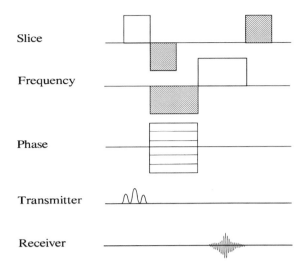

Slice

Frequency

Phase

Transmitter

Receiver

**Figure 200**  Gradient–echo imaging pulse sequence. The gradient pulse applied along the slice axis following acquisition of the gradient echo is used to destroy any coherent $M_{xy}$.

tion of $M_z$. This allows the sequence to be run with extremely short TR intervals. If small flip angles, rather than the 90° pulse, are used for excitation of the slice, less longitudinal magnetization is saturated, and even shorter TR values may be used. The gradient–echo sequence allows complete images to be acquired in seconds rather than the minute time scale required for standard spin–echo images.

The substantial time savings are a significant advantage of the gradient–echo pulse sequence. Other features of the gradient–echo sequence are both advantageous and deleterious. The quality of the images acquired with the gradient–echo sequence is sensitive to small variations in the static dc magnetic field. The process of gradient reversal only compensates for the phase shift induced by the applied bipolar magnetic-field gradient. However, small gradients are also caused by inherent inhomogeneities in the static dc magnetic field, as well as gradients originating within the sample due to regional differences in the magnetic susceptibility. Unlike the spin–echo sequence, the phase effects of these persistent gradients are not refocused by gradient reversal. When magnetic field gradients are present across an image voxel, the precession of the individual nuclei is out of phase causing destructive interference in the signal generated by these dipoles. In gradient–echo sequences these field perturbations produce a loss in signal intensity. Gradient–echo imaging sequences require a uniform static magnetic field. Certain pathologic processes, such as hemorrhage, alter the magnetic susceptibility of tissue, thereby generating small field gradients. Due to the sensitivity to magnetic-field variations, gradient–echo sequences may be more useful for detecting such lesions.

Contrast development in the gradient–echo image is slightly different from that described for spin–echo sequences. The parameters for gradient–echo imaging sequences are summarized in Table XVII. With the gradient–echo sequence, $T_1$-weighting can be adjusted by varying either the TR value or the flip angle of the excitation pulse. As larger flip angles are used, more longitudinal magnetization

**Table XVII**

Dominant parameters that affect image contrast in gradient–echo images

|  | Short TE | Long TE |
|---|---|---|
| Short TR | $T_1$, $T_2$(SSFP)[a] | $T_1$, $T_2$, $T_2^*$ |
| Long TR | Proton density | $T_2$, $T_2^{\,b}$ |
| Large flip angle | $T_1$, $T_2$(SSFP) | $T_1$, $T_2$, $T_2^*$, $T_2$(SSFP) |

[a] Increased steady-state free precession signal composed of signal from dipoles with longer $T_2$ values.

is excited and a longer TR period is required for relaxation. Thus for a constant TR value the sensitivity of the NMR signal to $T_1$ times of the sample increases as the flip angle increases. Analogous to this case, for a constant flip angle the amount of $T_1$-weighting increases as the TR value is shortened. As with previous sequences, increasing the TE increases the $T_2$-weighting of the sequence.

If TR values on the order of the $T_2$ of the sample are used, transverse magnetization persists between scans. Since this signal accumulates phase shifts induced by the phase-encoding gradient, it can interfere with the gradient–echo signal producing intensity artifacts in the image. To minimize these artifacts a gradient pulse is applied following data acquisition to destroy the phase coherence of the transverse magnetization. Alternatively the persistent signal can be used constructively. As diagrammed in Fig 201, reversing the phase gradients following acquisition of the echo refocuses the magnetization at the time of the next 90° excitation pulse. This establishes a steady-state free precession (SSFP) in which the transverse magnetizaton present prior to the excitation pulse contributes signal to the subsequent gradient echoes. Tissues with long $T_2$ values contribute more to

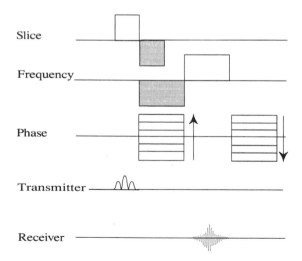

**Figure 201** Gradient–echo sequence used to obtain steady-state free precession signal. The direction of the phase-encoding gradient is reversed following acquisition to refocus the $M_{xy}$ prior to the next scan.

the steady-state free precession. Increasing the flip angle or decreasing the TR value generates more steady-state signal. Under conditions that establish steady-state precession this will increase both the $T_1$ and $T_2$ weighting of the sequence.

## D.  Blood-Flow Imaging

Imaging of blood flow in the human body consists of a variety of techniques used to visualize differences in vascular structures of interest. An example of blood-flow imaging or magnetic-resonance angiography is shown in Fig. 202. Parameters of blood-flow direction, velocity, vascular geometry, and the imaging parameters must be optimized to obtain information regarding the vascular structure of interest. Typically there are four clinically useful applications of blood-flow imaging: (1) 2-D time-of-flight, (2) 3-D time-of-flight, (3) 2-D phase contrast, and (4) 3-D phase contrast.

To understand blood-flow imaging techniques, a basic familiarity with hemodynamics is necessary. Flow resistance in a straight blood vessel is related to blood viscosity and length of the vessel and inversely related to the diameter of the vessel. The velocity profile of the flow across the vessel is approximately parabolic. This means that flow near the vascular wall is slower than that in the center of the vessel.

Flow becomes more complex near areas of stenosis (narrowing of the vessel). As blood travels through a stenosis the velocity abruptly increases and vortices begin to form just beyond the stenosis. Finally, the flow becomes turbulent and later returns to normal at some distance beyond the stenosis. Complex flow often results in decreased image signal intensity and must be distinguished from regions of abnormal pathology. The difficulty is that regions of atherosclerotic disease are

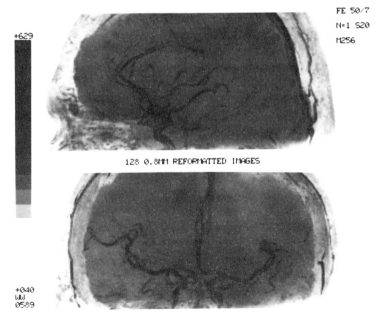

**Figure 202**  Magnetic resonance angiography (MRA). (Courtesy of Toshiba American Medical Systems, Tustin, California)

also present with decreased signal intensity. Thus blood-flow imaging techniques must be chosen and optimized to discriminate natural loss of signal and true disease expression.

Loss of signal in a blood-flow image can occur from either saturation of nuclear magnetization or from loss of coherence of the magnetization. The coherence can be disrupted when there are variations in the magnetic-field homogeneity either spatially or due to local magnetic susceptibility effects. Loss of coherence can also occur due to broad variations in flow velocities or for uncompensated motions of the sample. Variation in flow velocities is prevalent in both normal and diseased blood vessels. In normal vessels the parabolic variation in velocity is due to the increased shear force near the vessel's wall in normal laminar flow. Variation in diseased vessels can be caused by stenosis, aneurysm, or

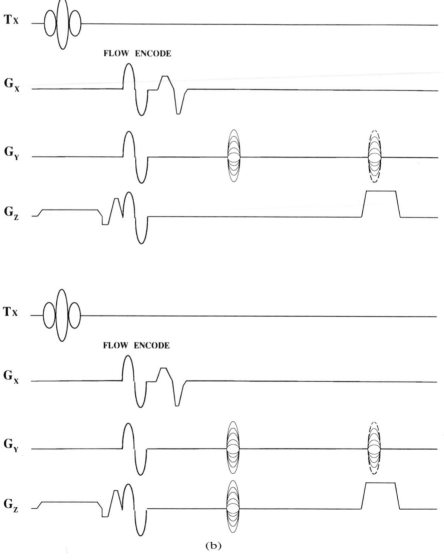

(b)

**Figure 203** Simplified diagram of (a) 2-D and (b) 3-D time-of flight pulse sequence.

any other disruption of normal flow. The spin–echo experiment is particularly good for sensing these kinds of signal loss since spin echoes refocus less effectively as coherence is disturbed.

In time-of-flight angiography the magnetization in a slice of the tissue sample perpendicular to the direction of flow is saturated with a radio-frequency pulse (Lenz *et al.*, 1988; Wehrli *et al.*, 1986; Axel, 1984; Dumoulin, 1989; Marchal *et al.*, 1990). A brief time delay allows unsaturated magnetization to enter the slice with normal blood flow. An image is quickly taken that results in a larger signal for the entering blood than for the stationary tissue. The pulse sequence for 2-D and 3-D time-of-flight imaging is shown in Fig. 203. Blood entering from either side of the slice will appear bright. Often an additional saturation pulse will be applied to a second slice either above or below the region of interest to selectively saturate unwanted venous flow from the image.

The phase-contrast method of blood-flow imaging differs from the time-of-flight techniques. Instead of using the unperturbed magnetization of the inflowing blood to produce the signal, it relies on the velocity-modulated loss of coherence (Spritzer *et al.*, 1990). The diagram in Fig. 204 shows the pulse sequence used for both 2-D and 3-D phase-contrast angiography. Once the magnetization has been tipped into the axis orthogonal to the static field, its precession changes the phase of the magnetization vector. When a bipolar gradient is applied it encodes the spin's velocity as a function of the changes of this phase angle. Motions of constant velocity will result in phase shifts, but absence of motion or nonuniform movements will not cause the phase to shift. Thus, uniform motion such as the flow of blood will be seen as a hyperintense volume, while random movements of the body and stationary tissue will be hypointense. Normally the bipolar gradient is applied in only one direction. In regions such as the head, where flow is present in almost all directions, flow can be evaluated in all three axes and reduced to flow magnitude independent of direction (Dumoulin *et al.*, 1989).

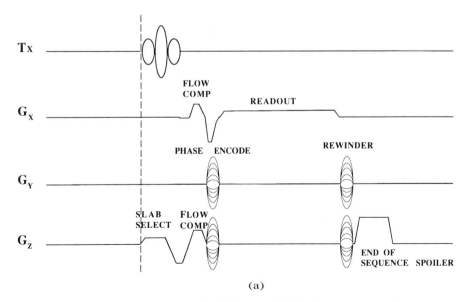

(a)

**Figure 204** Simplified diagram of (a) 2-D and (b) 3-D phase contrast pulse sequence. FLOW COMP is flow compensation.

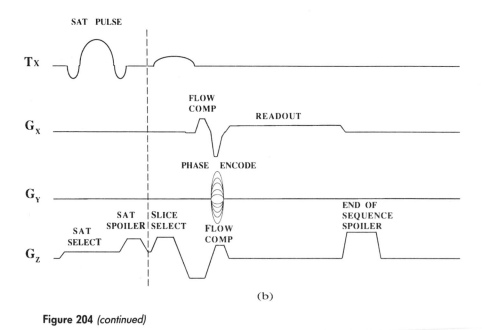

Tx

FLOW
COMP

$G_X$

READOUT

PHASE ENCODE

$G_Y$

SAT | SLICE
SPOILER | SELECT

END OF
SEQUENCE
SPOILER

SAT
SELECT

FLOW
COMP

$G_Z$

(b)

**Figure 204** *(continued)*

# IV. *in vivo* NMR Spectroscopy

Many techniques have been used to assess cellular energy metabolism in tissue and organs. NMR is a noninvasive method for characterizing anatomic structure and metabolic state.

## A. Historical Details

The application of NMR to living systems began shortly after its conception by Block and Purcell. Block is said to have detected a proton signal from water by placing his finger in the radio-frequency coil of the spectrometer (Gadian, 1982). Odeblad and Lindstrom demonstrated the early use of NMR in biomedicine by studying the signals from blood cells and a variety of excised tissues (Odeblad and Lindstrom, 1955). Their comments concerning signal intensity and relaxation have been hinted as the foundations of NMR imaging. Further, in 1973, Moon and Richards (1973) demonstrated how the chemical shift of certain resonances can be used in determining the intracellular pH using phosphorus ($^{31}$P) NMR. In 1974, Hoult *et al.* (1974) produced the first $^{31}$P spectra of intact muscle that contained resonances identified as adenosine triphosphate (ATP), inorganic phosphate ($P_i$), phosphocreatine (PCr), and sugar phosphate. Larger magnet size has led to the logical progression of the study of tissue and disease in humans.

## B. Chemical Shift and Identification

An early attraction of NMR was the identification of unique molecular species by chemical shift. Arnold *et al.* (1951) observed that elements of the same molecule possess slightly different resonance frequencies (Arnold *et al.*, 1951). The origin of this difference is the alteration of the magnetic field intensity due to the screening or shielding of electrons surrounding the nucleus (Fig. 205). The mo-

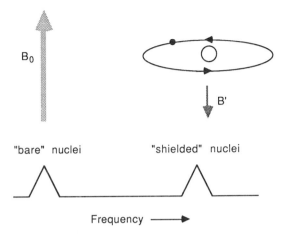

**Figure 205**   Schematic representation of the orbital motion of an electron around its nucleus induced by the static magnetic field $B_0$. The movement of this electron creates an opposing and smaller magnetic field $B'$ (top). The opposing magnetic field decreases the intensity of the total magnetic field experienced by the nucleus. Therefore, because the frequency of the nucleus is proportional to the total magnetic field strength, the frequency of the nucleus is reduced when shielded by electron motion (bottom). The decrease in resonance frequency is dependent on the local electron density around the nucleus. Changes of resonance frequency of a nucleus due to electron shielding are defined as chemical shift.

tion of the electrons produces a magnetic field that opposes the primary static field $B_0$ so that the nucleus experiences a local field intensity slightly less than $B_0$. Since the resonance frequency depends on the strength of $B_0$, each nucleus resonates at a unique frequency and reflects the electron density that surrounds it. These frequencies in hertz (Hz) are often normalized to the given magnetic field in units of parts per million (PPM).

## C.   Quantitation

The intensity of the emitted RF resonance signal from the excited sample is proportional to the concentration of a particular nucleus in the applied static magnetic field, and its position along the frequency axis is determined by the strength of the applied dc magnetic field and the degree of electron shielding around the nucleus itself. The width of the resonance line is affected by several parameters, which may include pH, spin–lattice ($T_1$) and spin–spin ($T_2$) relaxation, nuclear Overhauser enhancement (Overhauser, 1955), and chemical exchange. The use of surface coils complicates quantitation due to the inhomogeneous detection of the sample volume. However, methods are available to overcome these difficulties.

In the application of high-resolution NMR to tissue and cells, only the mobile molecules are observed. The signals from immobile molecules are present, but are often too broad to be observed due to enhanced $T_2$ relaxation. Thus, in general, signals from macromolecules such as membrane phospholipids, DNA, and RNA are not seen or seen as very broad components of the resonance line. A good example of signal from immobile structures in $^{31}P$ NMR is the signal from bone, although broad, can be seen due to its high phosphorus concentration.

The relaxation processes, $T_1$ and $T_2$, have been previously discussed. To accurately quantitate the concentrations of metabolites, nuclei must be allowed to relax completely, that is, to return to their lowest energy state before the next radiofrequency pulse is applied. When accumulated in this way, the intensities or the areas of the individual resonances are proportional to the total number of nuclei within the sample.

## D.  Sensitivity and Resolution

One of the major disadvantages of the NMR technique is its low sensitivity. The observation of metabolites *in vivo* is presently limited to concentrations of roughly $10^{-3}$ M or higher. The current level of sensitivity has been enhanced to its present state by the use of Fourier transform techniques and the development of superconducting magnets. An obvious way to increase sensitivity is to increase the signal-to-noise ratio by increasing the applied magnetic-field strength since the signal/noise would be reasonably proportional to $B_0^{7/5}$. There is clearly both a structural and possible safety limitation to further increases of the static field for whole-body magnets. The lower sensitivity is occasionally fortuitous since the observable nuclei are usually limited to small mobile molecules of high concentration. This results in the simplification of spectra due to fewer numbers of compounds present and often less overlap between resonance lines.

## E.  pH and Metal Ion Binding

One of the more important aspects of NMR *in vivo* has been in its use for the evaluation of intracellular pH (Fig. 206). The technique exploits the chemical shift dependence due to fast exchange of such molecules as inorganic phosphate or change in pH. The chemical shift arising from inorganic phosphate and certain phosphomonoesters (i.e., glucose 6-phosphate) is used since their $pK_a$ values are in the range of physiological pH. The $pK_a$ is defined as the negative log of the chemical association constant. In unusual circumstances, the chemical shift of the gamma phosphate of ATP is used in tissues or organelles possessing more acidic pH. The method is routinely calibrated by titration of the appropriate phosphorus compound in an environment similar to that of the tissue of interest.

At pH = 7 inorganic phosphate is present in two forms as $(HPO_4)^{-2}$ and $(H_2PO_4)^{-1}$. In the absence of chemical exchange processes, two individual resonances would be seen corresponding to the two forms at frequencies $f_b$ (basic) and $f_a$ (acid). However, the two forms of inorganic phosphate are in fast chemical exchange and only a single resonance results. The position or frequency of this resonance is determined by the quantity of each of the two forms of phosphate. Therefore, the pH is predicted by the Henderson–Hasselbach equation as

$$pH = pK_a + \log[(f - f_a)/(f_b - f)]$$

where $f$ is the measured NMR frequency of the single inorganic phosphate resonance.

Errors can occur from an improper evaluation of the constituency of the cellular fluid. Extensive investigation has been performed on factors other than pH that influence local shift behavior. The resonance frequencies of phosphates have been shown to be altered by interactions with proteins, phospholipids, and nucleic acids as well as metal ions such as magnesium, potassium, and calcium (Roberts *et al.*, 1981). Large frequency shifts are also observed with changes in ionic

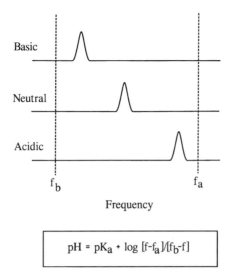

$$pH = pK_a + \log [f\text{-}f_a]/[f_b\text{-}f]$$

**Figure 206**  Measurements of pH by $^{31}$P NMR spectroscopy. At pH = 7, inorganic phosphate is present in two forms as $(HPO_4)^{-2}$ and $(H_2PO_4)^{-1}$. In the absence of chemical exchange processes, two individual resonances would be observed corresponding to the two phosphate forms at frequencies $f_a$ and $f_b$. The species are in fast chemical exchange, however, and only a single resonance is produced. The position or frequency (*f*) of the resonance is determined by the concentration of each of the two forms of inorganic phosphate. Therefore, pH is predicted by the Henderson–Hasselbach equation, where *f* is the measured frequency of inorganic phosphate. The figure shows the position of the inorganic phosphorus resonance at basic, neutral, and acidic pH.

strength (Roberts *et al.*, 1981). The measurement inaccuracies of the pH may thus be in the order from 0.2 to 0.5 pH units or higher, if cellular ionic strength and environment are not well defined (Gadian, 1982). Consequently, care is required in the application of this technique.

## F. Measurement of Energy Metabolism

Adequate concentrations of ATP are central in providing the necessary energy to maintain metabolite levels, ion pumps, and other cellular functions. Decrease of energy supply or oxygen will result in failure of the cellular ion and metabolite regulation. NMR is well suited to monitor noninvasively the phosphorus energy potential at a time resolution of several minutes.

In most tissues the concentration of phosphorus-containing metabolites such as PCr, ATP, $P_i$, monophosphoesters (ie., sugar phosphates), and diphosphoesters (ie., phospholipids) can be measured. Additionally the pH can be obtained, as previously discussed, by the frequency difference between the resonances for PCr and $P_i$. Ratio of metabolite concentration such as [PCr]/[$P_i$] are used to describe cellular health. This ratio is not only a sensitive criterion, but is based on standard measures of biochemical energy reserve. Let us consider the calculation of the phosphorylation potential ($\Gamma$).

$$\Gamma = [ATP]/[ADP][P_i]$$

This quantity and the pH determine the free energy of hydrolysis of the terminal phosphorus bond of ATP. It measures the energy available to the cell to perform work, indicating the ratio of the higher energy substrate to its lower energy products. The value of $\Gamma$ is difficult to measure in the cytosol due in part to the difficulty in measuring ADP. A significant portion of the total ADP of the cell is often protein bound.

The concentration of ADP can be evaluated using the steady-state equilibrium constant ($K_{eq}$) of the creatine kinase reaction

$$PCr + ADP + H^+ \rightleftharpoons Cr + ATP$$

Combining the equations above, the NMR measurable parameter of [PCr]/[P_i] can be shown to vary proportionally to $\Gamma$ as given by

$$[PCr]/[P_i] = (\Gamma[Cr])/(K_{eq}[H^+])$$

In this equation, total [Cr] can be estimated from biochemical analysis and [H$^+$] is known from the measurement of the pH. The [PCr]/[P_i] ratio provides a sensitive monitor of cellular energy supply since PCr is the immediate source used to maintain ATP of the tissue. Results are often expressed as the [PCr]/{[PCr] + [P_i]} ratio, which normalizes the ratio and avoids extreme values of zero and infinity of [PCr]/[P_i] due to near-noise-level values of PCr or P_i.

Finally, the variation in the [PCr]/[P_i] can be shown to vary with the velocity of reaction ($v$) for the Michaelis Menton kinetics of the creatine kinase enzyme (Chance *et al.*, 1988). This equation in terms of NMR-measurable parameters is ($v_{max}$ is maximal velocity reaction)

$$v = \frac{v_{max}}{\dfrac{1 + 8.1 \times 10^{-3}/[Cr]}{\{[P_i]/[PCr]\}}}$$

# V.   Characteristics of Magnetic Resonance Images

## A.   Spatial Resolution

The spatial resolution of the MR image is characterized by the number of pixels. The number of pixels is defined by the product of number of gradient steps along the phase-encoded and read axes. This matrix of pixels is often seen defined as $64 \times 64$, $128 \times 128$, $256 \times 256$, etc. The spatial resolution depends on the limits of signal sensitivity of a pixel's volume. The "depth" of the pixel is determined by the slice selection thickness. Increasing the viewing plane resolution by a factor of two will decrease the signal volume by a factor of 4. Thus, resolution is improved only at a significant expense to image signal-to-noise ratio.

The instrument's ultimate resolution can be limited by patient movement, poor gradient performance, etc. Resolution may be masked when affected by other factors such as chemical shift artifacts. Water and fat protons are both present in the human body, but resonate at slightly different chemical shift or frequency. This results in the smearing of signal intensity into spatial pixel locations in which the signal does not originate.

## B.  Image Contrast

The three principal tissue parameters that determine signal intensity are $T_1$, $T_2$, and proton density. Some limited contrast exists due to proton density or the number of hydrogen atoms found in a particular tissue volume. However most contrast is generated by exploiting difference in signal intensity due to $T_1$ and $T_2$ relaxation found at selected intervals of TE and TR.

Image contrast can be manipulated by other techniques. Inversion recovery imaging is a good example. During an inversion recovery sequence the magnetization is inverted, the magnetization is allowed to relax for a variable time period, and then magnetization is read by a final 90° pulse. The variable time period can be adjusted to the exact time at which the magnetization passes from a negative to a positive signal. At this time no signal will be detected. If an inversion recovery image is made of a tissue with at least two distinct $T_1$ values, the variable time can be adjusted to diminish the signal of one tissue type relative to the other, creating contrast.

A slightly more elaborate contrast technique is magnetization transfer imaging (Wolf and Balaban, 1988). In this scheme a normal image is taken, but the sample is irradiated with a narrow bandwidth of frequency, many kilohertz from the central water resonance. Very broadened magnetization due to the interaction of water with large proteins is saturated by the irradiation. Saturation refers to a process in which the populations of nuclear spins become equal, making a signal impossible. These saturated molecules are in exchange with the bulk water of the sample. Net transfer of magnetization from the narrow water resonance decreases its intensity. If this process occurs more favorably in one tissue type than another, contrast is produced.

# VI.  Biological Effects
# of Magnetic Fields

## A.  Static Magnetic Fields

### 1.  Introduction

Interest in the biological effects of magnetic fields dates back to the early Greeks and Chinese (Stoner, 1972; Needham, 1962; Carlson, 1975; Malmstrom, 1976). Magnetic ores were mined from the Greek province of Magnesia in Asia Minor and hence given the name magnet or magnetite. Early investigators believed that magnets were "alive" and possessed animating or magical powers. In the early 1770s some attempts were made to use magnets or lodestones to treat human disease (Darnton, 1970) but were soon dismissed as fraudulent.

More recently magnetic properties and spectroscopy of biological materials have been an active area of study. Specifically for NMR methods, fields on the order of 0.05 to 4.0 tesla and gradients averaging 0.01 T/m are commonly used. Rapid switching of gradients can involve changes in magnetic fields of as much as 2 T/sec. Due to the impact of MRI on clinical medicine it is a common practice to have humans exposed to static magnetic fields at 0.5 T or higher.

## 2. Interaction of Static Magnetic Fields with Living Tissue

Most of the human body is diamagnetic although areas of paramagnetic character exist, usually due to the presence of iron in the specific tissues or blood. In general, however, the relative magnetic permeability ($\mu_r$) of the body is close to unity. Thus, the values of the magnetic field ($H$) and the magnetic induction ($B$) are related by the following expression

$$B = \mu_r \mu_0 H = (1 + X_m)H$$

where $X_m$ is the relative magnetic susceptibility. In regions of the tissue where the $\mu_r$ or $X_m$ is higher, the magnetic field induction ($B$) will become higher. Magnetic field variation in the local environments of the cell may play some role in the metabolic rate of normal cells. Although the data in this area are sometimes contradictory, often chemical reactions are accelerated in magnetic fields by higher susceptibility and decrease with diamagnetic susceptibility. Accelerated rates are commonly seen in reactions involving iron centers, such as catalase, while synthesis reactions (RNA and DNA), which occur largely without iron centers, decrease (Gaffney and McConnell, 1974; Gaffey and Tenforde, 1981; Hirschbein *et al.*, 1982; Audus, 1960; Audus and Whish, 1964). Magnetic fields of $10^{-3}$ to $10^{-2}$ tesla have been shown to affect chemical reactions by influencing the electronic spin state of the reaction intermediates (Atkins, 1976; Bube *et al.*, 1978).

Alignment of molecules has been demonstrated in intense magnetic fields where enough diamagnetic anisotropy exists. Muscle fibers (Arnold, *et al.*, 1958;), chloroplasts (Geacintov *et al.*, 1971), retinal elements (Chagneux *et al.*, 1977; Chambre, 1978), sickled erythrocytes (Murayama, 1965), membranes (Hong, 1977; Neugebauer *et al.*, 1977), and other macromolecules (Maret and Dransfield, 1977) possess adequate diamagnetic anisotropy to produce highly oriented structure. In plants these magnetically induced alignments have been implicated in redistribution of membrane proteins that regulate intracellular $Ca^{+2}$ concentration (Sperber *et al.*, 1981). Magnetic ordering of phospholipid bilayers (Gaffney and McConnell, 1974) and magnetically induced changes in fluidity of gels due to alignment of agarose monomers (Kalkwarf and Langford, 1979) have been demonstrated.

## 3. Field Interactions with Moving Conducting Volumes (Heart and Blood Flow Effects)

It is well known that moving a conductor in a magnetic field will produce a current in the conductor. This can be seen from Ohm's law

$$\mathbf{I} = \sigma \mathbf{E} = \sigma (\mathbf{v} \times \mathbf{B})$$

where $\sigma$ is conductivity and $\mathbf{v}$ is the velocity of the conductor movement and $\mathbf{E}$ is the electric field. Further, a net volume force ($F$) is produced on the conducting volume defined by

$$\mathbf{F} = \mathbf{I} \times \mathbf{B}$$

This force is generally small. At 2.0 tesla in the human aorta a decrease in axial flow of approximately 1% would be expected due to this force (Polk and Postow, 1986).

The more important consideration is the potential produced due to the pulsing or flowing blood. This potential is orientation dependent since it will be produced only when the movement of blood is orthogonal to the direction of the applied magnetic field. Major blood vessels and the heart are the only structures large enough for concern. Aortic induced potentials at 2.0 T have been estimated to be 4.71 mV (Neugebauer *et al.*, 1977). The potential across the individual heart cells is thought to be much lower and certainly lower than the 40 mV needed to depolarize the cells. This increase in potential can often be seen as an increase in the ECG waveforms to the T-wave amplitude. Experiments measuring electrical potential in squirrel monkeys have been performed in fields as high as 10 T. Animals exposed for 15 min exhibited no change in cardiac rate, but increased their T-wave amplitudes by 1.0 mV (Torbert and Maret, 1979). Longer exposures at as high as 10 T showed no changes in either breathing rate or the R-wave form, but did result in a decrease in heart rate and an increase in sinus arrhythmia and amplitude of the T-wave (Williamson and Kaufman, 1980; Erne *et al.*, 1982). Currently, a National Radiological Protection Board (NRPB) advisory suggests that exposure of individuals in NMR examination be limited to 2.5 T. This field strength is near the calculated value that can produce threshold potentials required to depolarize a myocardial cell. It should be noted that NMR examinations at 3.0 and 4.0 T are now routinely performed, so far without incidences, in several centers worldwide. Thus, the real limits of safe static field exposure for NMR-induced flow potentials remain to be resolved.

## 4. Effects of Static Magnetic Fields in Humans

Humans spend their entire lives in a "sea" of weak magnetic fields. Depending on location, the earth's magnetic field ranges from 0.2 to 0.5 gauss. Increases in field due to heavy iron-containing building materials, even our automobiles, increase our static field local exposure to many times this value. It is difficult therefore to study the effects of static field under completely controlled condition in humans.

Earliest records of human exposure took place in the Edison Laboratories in 1892 (Peterson and Kennelly, 1892). In this study a boy was exposed to magnetic fields of 0.1 to 0.2 T for an undisclosed period of time. No adverse effects were recorded. The development of the nuclear accelerator precipitated the first routine long-term exposure to high static magnetic field in humans. Again, no adverse reports are recorded. The reports of effect of fields as high as 10 T are anecdotal, but have remained consistent over the years. Complaints of a feeling of bitter cold, aches, pain, and the feeling of ants crawling on the extremities are common (Ketchen *et al.*, 1978). At fields of 1.5 T individuals with metallic fillings reported a strange taste and pain in the filled teeth (Ketchen *et al.*, 1978). It should be noted that this pain may be due to the force produced on all nonmagnetic metals as they pass through areas of high magnetic flux, and not with the magnetic field directly.

The first serious study to look at static magnetic field effects involved industrial workers whose hands and heads were exposed to fields of 0.035 to 0.35 T and 0.015 to 0.15 T, respectively. These individuals reported symptoms of headaches, pain near the heart, dizziness, distorted vision, fatigue, insomnia, loss of appetite, increased perspiration, and itching of the hands and wrist (Vyalov, 1971). A small decrease in blood pressure in roughly a third of the workers was reported. Many of these symptoms could not be disassociated from the work environment, which subjected the worker to high temperatures, metallic dust, emulsions, and degreasing solvents.

A more recent study conducted on aluminum reduction plant workers who were exposed to fields of 0.0005 to 2.0 T showed no significant increase or decrease in prevalence of disease (Rockette and Arena, 1983). This study was particularly well controlled for the time, considering age, race, and socioeconomic class.

In conclusion, the existing evidence is still incomplete for static field exposure in both animals and humans. Larger and better-controlled studies will be needed to resolve the question. Several conclusions can presently be made. There is no evidence to suggest danger from mutagenic effects, blood-flow induced potentials, or organ development for humans in fields normally used for clinical MRI examinations. As field strength for human examination increase, caution should be exercised as flow-induced potential becomes larger. Still the most significant danger remains the presence of ferromagnetic objects internal or external to a patient in high magnetic fields. Cumulative exposure in patients and in workers in MRI facilities have not resulted in any reports of damaging effects.

## B.  Radio-Frequency Fields

During *in vivo* imaging or spectroscopy examinations only a small amount of the transient high-power RF pulse is absorbed by the nuclei in the tissue. The majority of this energy is rapidly converted to heat the tissue. The efficient transfer of this energy to the tissue is determined by the frequency, pulse amplitude and duration of the RF energy; conductivity and dielectric constant of the tissue; and proximity and geometry of the RF transmitter. Ultimately, even the efficacy of

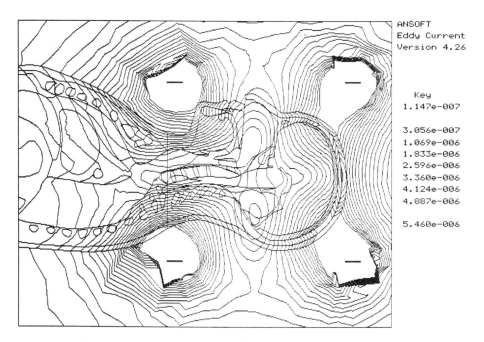

**Figure 207**    Simulation of the radio-frequency field distribution calculated at 200 MHz in the head and upper body. Distortion of the field is observed due in part to self-resonance of the head and neck.

heat removal by tissue perfusion is important to consider. The useful quantity of specific absorption rate (SAR) to describe heat accumulation is defined as

$$SAR = \sigma E^2 D_c / 2\eta$$

where $\sigma$ = tissue conductivity, $E/2$ is the average induced electric field, $D_c$ is the duty cycle or ratio of transmitter pulse to repetition interval, and $\eta$ is the tissue density. In tissues receiving perfusion, the body dissipates energy continually from its basal metabolism of about 1 W/kg. SARs near this value should not pose significant difficulties for tissue. This may not be true for tissues with limited or no blood flows, such as the testes and lens of the eye. The absorption of radio-frequency energy in humans is further complicated since energy is not distributed evenly over the body. This occurs primarily because of self resonances of body parts that are related to the size, conductivity, and dielectric constant of the tissue. In humans, increases in energy depositions occur in the legs and arms at 47 MHz (Brune and Edling, 1989). Genuine partial body resonance occurs in the head and neck at 100–400 MHz (Brune and Edling, 1989). Standing RF waves can be produced in the head which may result in "hot spots" at frequencies of 200–400 MHz (Smith, 1991). A calculation is shown in Fig. 207 of the standing RF waves occurring in a simulated adult head at 200 MHz.

## C. Linear-Gradient Magnetic Fields (Extremely Low Frequency Fields)

### 1. Introduction

The MRI technique uses a magnetic-field gradient to spatially encode the protons within a sample under view. These pulsed linear gradients usually have intervals of 10 to 100 msec and are repeated every 20 to 2000 $\mu$sec. By definition time-varying magnetic fields below a frequency of 300 Hz are extremely low frequency (ELF) magnetic fields. There is a significant interest in ELF fields since they operate at a frequency that includes the natural working frequency of the brain and nerve conduction. It is thought that an ELF field can produce electrical currents that can interfere with the function of the neuron (Gaffey and Tenforde, 1981; Gross and Smith, 1961; Gualtierotti, 1964).

### 2. Magnetophosphenes

D'Arsonval reported in 1893 the visual sensation of flashes of light caused by ELF fields (Barnothy, 1963). These flashes of light were linked to the induced formation of magnetophosphenes in the eye (Beischer, 1964; Beischer, 1962; Frazier, 1980; Ketchen et al., 1978). Estimates of the threshold field needed to precipitate this effect range from 1.3 to 5.0 T/sec (Beischer, 1969). This phenomenon is frequency dependent with an optimum ELF magnetic field near 20 Hz. Induction of magnetophosphenes has not only occurred with pulsed MRI, but with head movement in static magnetic fields rapid enough to produce the threshold field changes. This induction will sometimes develop, although rarely, when patients are placed into the bore of the high-field MRI magnet.

### 3. Bone Healing

Pulsed ELF magnetic fields have been shown to increase both healing and growth in bone. Application of low frequency (65 Hz) and intensity (20 mV/cm) pulsed

ELF magnetic fields significantly improved the healing of fractured canine fibulae (Barnothy, 1964). The growth rate of embryonic chicken tibiae increases with application of pulsed electric fields (Leusden, 1929). Fitton-Jackson and Bassett (Chalazonitis and Arranitake, 1962) have demonstrated that calcium uptake and collagen content are both increased in developing chicken bone with pulsed ELF magnetic field exposure. Whether the piezoelectrical properties of bone are involved in this behavior is unknown. Eighty-seven percent of united fractures of the tibial diaphysis in humans have shown improved healing using pulsed electromagnetic fields (Barnothy, 1964). The technique is currently receiving serious orthopedic consideration.

## 4. Heart and Muscle Stimulation

The myofibrils of the muscle are stimulated to contract by electrical currents from the nerves. Applied currents can override the normal neuronal control and cause the muscles to contract. This can have a particularly dangerous effect in heart muscle if the currents are high enough to depolarize the myocardial cells, possibly causing ventricular fibrillation. Typical currents of 300 $\mu$A/cm$^2$ are required to induce heart fibrillation (Haberdutzl, 1967; Halpern and Green, 1964; Kolin, 1952). It is expected that ELF field changes of 500 T/sec (roughly 500 $\mu$A/cm$^2$) would be required for the induction of fibrillation (Halpern and Green, 1964). This can be achieved by pulsing a small ELF field very rapidly. The action of the ELF field on the muscle is dependent on the waveform, frequency, amplitude, and duration of the field exposure. Generally, the heart should be able to safely absorb a sinusoidal field with an amplitude of 5 mT at a frequency of 20 Hz (Persson and Stahlberg, 1989). These amplitudes and frequencies are similar to experimental conditions used for echo-planar imaging. Reports of stimulated skeletal muscle during echo-planar imaging at gradient strengths of 50 mT/cm have been observed (Smith, 1991).

Slightly larger current densities 3000 $\mu$A/cm$^2$ are required to induce seizures (Brown and Skow, 1978). Reversible nerve damage has been observed to occur in animals with 1,000,000 $\mu$A/cm$^2$ (Beischer, 1964). Fortunately currents of this magnitude are normally not reached in head MRI examinations. Additionally the amplitude of the pulsed ELF field is spatially distributed during MRI examination. Typically the linear gradients have a value of zero at the isocenter of the magnet. Thus, the higher ELF magnetic-field amplitude will be produced some distance from the isocenter. Therefore, in a standard imaging scheme the muscles in the shoulder and arms might be the first to be stimulated.

## D. Imaging Safety

All the preceding magnetic field effects should be carefully considered in exposing patients to new field strengths or protocols. A few areas, however, have become routine safety procedures. The most common risks are still those associated with metal objects inside or on the individuals within or near the magnet. Screening carefully for small ferromagnetic objects are among the most routine problem faced by clinical facilities. Equally important are implanted metal objects which will experience a strong force due to the field during scanning. The object need not be large if it is associated with delicate tissues that might tear as a result of torque produced by the static field. Implanted metal in the right configuration can occasionally couple to the RF magnetic field and produce heat. The NRPB has recommended that RF exposure should not increase the core body temperature by more than one degree. Irradiations less than 0.4 W/kg for the whole body or 4 W/kg for any 1 g of tissue are good guidelines. Larger magnets, higher frequencies, rapid imaging modalities, and regional saturation are constantly pressing closer to these limits.

## Questions

### True or False

1. The $T_2$ of a tissue is always greater than or equal to the $T_1$ of the tissue.
2. A gradient–echo imaging sequence is more sensitive to static magnetic field inhomogeneities than a spin–echo imaging sequence.
3. MRI has better soft tissue contrast than CT.
4. MRI is insensitive to motion.
5. The NMR signal does not penetrate through bone or air.
6. Slice-selection requires the simultaneous application of RF energy and a linear magnetic-field gradient.
7. The frequency-encoding gradient introduces phase artifacts into the signal that are corrected using a dephasing gradient prior to the 180° refocusing pulse.
8. The inversion recovery sequence is used to maximize contrast due to differences in the $T_2$ of tissues.
9. At equilibrium the net transverse magnetization ($M_{xy}$) is equal to 0.
10. For a selective RF pulse the pulse duration (PW) is directly proportional to the frequency bandwidth of the pulse.

### Multiple Choice

1. The time domain signal is converted to a frequency domain signal through the use of a
   - **a.** Hilbert transform
   - **b.** Back projection algorithm
   - **c.** Fourier transform
   - **d.** Digital receiver

2. Current Clinical MR imagers utilize
   - **a.** Superconducting magnets
   - **b.** RF surface coils
   - **c.** Digital waveform memory
   - **d.** All of the above

3. The time between successive 90° pulses in a gradient–echo imaging sequence is known as
   - **a.** TR
   - **b.** $T_1$
   - **c.** TI
   - **d.** TE

4. The time the spins are present in the *XY* plane prior to acquisition of the signal is known as
   - **a.** TR
   - **b.** $T_1$
   - **c.** TI
   - **d.** TE

5. The signal produced by a 180° refocusing pulse is known as the
   - **a.** Inversion signal
   - **b.** Spin echo
   - **c.** Gradient echo
   - **d.** Stimulated echo

6. Increasing the TR
   - **a.** Increases the $T_1$-weighting
   - **b.** Decreases the $T_1$-weighting
   - **c.** Increases the $T_2$-weighting
   - **d.** Decreases the $T_2$-weighting

7. Increasing the TE
   - **a.** Increases the $T_1$-weighting
   - **b.** Decreases the $T_1$-weighting
   - **c.** Increases the $T_2$-weighting
   - **d.** Decreases the $T_2$-weighting

8. Increasing the TI beyond the null point
   - **a.** Increases the $T_1$-weighting
   - **b.** Decreases the $T_1$-weighting
   - **c.** Increases the $T_2$-weighting
   - **d.** Decreases the $T_2$-weighting

9. Advantage(s) of the gradient–echo imaging sequence include
   - **a.** Less sensitivity to magnetic field inhomogeneity
   - **b.** Greater $T_2$ contrast
   - **c.** Faster imaging times
   - **d.** Higher gradient currents

10. Increasing the $B_0$ field strength
   - **a.** Increases the Larmor frequency
   - **b.** Increases the magnet bore size
   - **c.** Increases the $B_1$ field strength
   - **d.** Increases image resolution

## Problems

1. For a 6 G/cm slice-selection gradient, what pulse width should be used to excite a 4 mm thick slice?
2. What is the Larmor frequency for a proton in an external field of 1.5 T? For a $^{31}$P nucleus?
3. What is the total imaging time for a spin–echo sequence with a 256 × 256 image matrix, a TR of 2 sec, and a TE of 70 msec?
4. What gradient strength is needed to generate a frequency bandwidth of 500 Hz over a distance of 1 m?
5. What will be the frequency change of a $^1$H in a red blood cell after 1 sec if it is moving parallel with a 6 G/cm magnetic field gradient at a velocity of 0.5 cm/sec?
6. A single scan has an S/N of 4.2. What is the S/N for 256 scans?
7. How many eigenstates are present for a $^{23}$Na nucleus?
8. What sampling frequency must be used to eliminate aliasing of a signal generated from a 20-cm sample in the presence of a 5-G/cm magnetic field gradient? (Do not assume quadrature detection, even if we have discussed it)
9. What is the bandwidth of a 2-msec Gaussian pulse?
10. What is the frequency bandwidth for a single pixel in a 256 × 256 image of a 20-cm sample, using a 5-G/cm frequency-encoding gradient?
11. A static magnetic field in an NMR imaging system has an uniformity of 5 kG ± 0.02 G. A field gradient of 0.4 G/cm is applied in the same direction. What is the difference in spatial frequency when two volumes of water are placed 1 cm apart? Calculate theoretical limit of the spatial resolution of the system.

## References and Further Reading

Arnold, J. T., Dharmitti, S. S., Packard, M. E. (1951). Chemical effects on nuclear-induction signals from organic compounds. *J. Chem. Physiol.* **19**, 507.

Arnold, W., Steele, R., and Mueller, H. (1958). On the magnetic asymmetry of muscle fibers. *Proc. Natl. Acad. Sci.* **44**, 1.

Atkins, P. W. (1976). Magnetic field effects. *Chem. Br.* **12**, 214.

Audus, L. J. (1960). Magnetotropism: a new plant-growth response. *Nature* **185**, 132.

Audus, L. J., and Whish, J. C. (1970). "Magnetotropism, in Biological Effects of Magnetic Fields." (M. F. Barnothy, ed.), 170. Plenum Press, New York.

Axel, L. (1984). Blood flow effects in magnetic resonance imaging. *AJR* **143**, 1157–1166.

Bangert, V., and Mansfield, P. M. (1982). *J. Phys. E. Sci. Inst.* **15**, 235.

Barnothy, M. (1963). Biological effects of magnetic fields on small mammals. *Biomed. Sci. Instrum.* **1**, 127.

Barnothy, J. M. (1964). Development of young mice. *In* "Biological Effects of Magnetic Fields, Vol. 1." (M. F. Barnothy, ed.) 93. Plenum Press, New York.

Barnothy, J. M. (1964). Rejection of transplanted tumours in mice. *In* "Biological Effects of Magnetic Fields, Vol. 1." (M. F. Barnothy, ed.) 1. Plenum Press, New York.

Beischer, D. E. (1964). Survival of animals in magnetic fields of 140,000 Oe. *In* "Biological Effects of Magnetic Fields, Vol. 1." (M. F. Barnothy, ed.) Plenum Press, New York.

Beischer, D. E. (1969). Vectocardiogram and aortic blood flow of squirrel monkeys (Saimire sciureus) in a strong superconductive magnet. *In* "Biological Effects of Magnetic Fields, Vol. 2" (M. F. Barnothy ed.) 241. Plenum Press, New York.

Beischer, D. E. (1962). Human tolerance to magnetic fields. *Astronautics* **7,** 24.

Bloch, F., Hansen, W. W., and Packard, M. E. (1946). *Phys. Rev.* **69,** 127.

Brown, F. A., Jr., and Skow, K. M. (1978). Magnetic induction of a circadian cycle in hamsters. *J. Interdiscipl. Cycle Res.* **9,** 137.

Brune, D. K., and Edling, C. (1989). "Occupational Hazards in the Health Professions," pp. 193–203 and 165. CRC Press, Inc., Boca Raton, Florida.

Bube, W., Haberkorn, R., and Michel-Beyerle, M. E. (1998). Magnetic field and isotope effects induced by hyperfine interactions in a steady-state photochemical experiment. *J. Am. Chem. Soc.* **100,** 5953.

Budinger, T. F., and Lauterbur, P. C. (1984). *Sci.* **226,** 288.

Bushong, S. C. (1988). "Magnetic Resonance Imaging: Physical and Biological Principles." C. V. Mosby, St. Louis.

Carlson, J. B. (1975). Lodestone compass: Chinese or Olmec primacy? *Science* **189,** 753.

Chagneux, R., Chagneux, H., and Chalazonitis, N. (1977). Decrease in magnetic anisotropy of external segments of the retinal rods after a total photolysis. *Biophys. J.* **18,** 125.

Chalazonitis, N., and Arranitake, A. (1962). Effets du champs magnetique constant sur l'autoactive des fibres myocardiques. *C. R. Seances Soc. Biol. Paris* **10,** 1962.

Chambre, M. (1978). Diamagnetic anisotropy and orientation of alpha-helix in frog rhodopsin and meta II intermediate. *Proc. Natl. Acad. Sci.* **75,** 5471.

Chance, B., Leigh, J. S., Jr., McLaughlin, A. C., Schnall, M., and Sinnwell, T. (1988). Phosphorus-31 spectroscopy and imaging. *In* Partain, C. L., Price, R. R., Patton, J. A. et al: "Magnetic Resonance Imaging, Vol. 2." W. B. Saunders, Philadelphia.

Cohen, S. M., and Shulman, R. G. (1980). *Phil. Trans. R. Soc. London B.* **289,** 407.

Crooks, L. E. (1985). *IEEE Eng. Med. Biol. Mag.* **4,** 8.

Curry, T. S., Dowden, J. E., and Murry, R. C., Jr. (1990). "Christensen's Introduction to the Physics of Diagnostic Radiology." 4th ed. Lea & Fehiger, Philadelphia.

Damadian, R. (1971). *Sci.* **171,** 1151.

Darton, R. (1970). "Mesmerism." Sckocken, New York.

Dumoulin, C. L. (1989). Magnetic resonance angiography, *Persp. Radiol.* **2,** 1.

Dumoulin, C. L., Souza, S. P., Walker, M. F., and Wagle, W. (1989). Three-dimensional phase contrast angiography. *Magn. Reson. Med.* **9,** 139–149.

Erne, S. N., Hahlebohm, H. D., and Lubbig, H., eds. (1982). "Biomagnetism." de Gruter, Berlin.

Everett, J. E., and Osemeikhian, J. E. (1963). *J. Sci. Instrum.* **43,** 470.

Frazier, M. E. (1980). Biological effects of magnetic fields, progress report. *In* "Health Implications of New Energy Technologies." (W. N. Rem and V. E. Archer, eds.) 679. Ann Arbor Science Publishers, Ann Arbor, Michigan.

Gadian, D. G. (1982) *In* "Nuclear Magnetic Resonance and its Application To Living Systems." Clarendon Press, Oxford.

Gaffey, C. T., and Tenforde, T. S. (1981). Alterations in the rat electrocardiogram induced by stationary magnetic fields. *Bioelectromagnetics* **2,** 357.

Gaffney, B. J., and McConnell, H. M. (1974). Effect of magnetic field on phospholipid membranes. *Chem. Phys. Lett.* **24,** 310.

Geacintov, N. E., van Nostrand, F., Pope, M., and Tinkel, J. B. (1971). Magnetic field effects on the chlorophyll fluorescence in Chlorella. *Biochim. Biophys. Acta* **226**, 486.

Gross, L., and Smith, L. W. (1961). Effect of magnetic fields on wound healing in mice. *Fed. Proc. Fed. Am. Soc. Exp. Biol* **20**, 164a.

Gualtierotti, T. (1964). Decrease of the sodium pump activity in the frog skin in a steady magnetic field. *Physiologist* **7**, 150.

Haberdutzl, W. (1967). Enzyme activity in high magnetic fields. *Nature* **213**, 72.

Halpern, M. H., and Green, A. E. (1964). Effects of magnetic fields on the growth of HeLa cells in tissue culture. *Nature* **202**, 717.

Hatfield, W. E., and Miller, J. E., Jr. (1988). "High Temperature Superconducting Materials." Marcel Dekker, New York.

Hirschbein, B. L., Brown, D. W., and Whitesides, G. M. (1982). Magnetic separations in chemistry and biochemistry. *Chem. Tech.*, March, **172.**

Hong, F. T. (1977). Photoelectric and magneto-orientation effects in pigmented biologic membranes. *J. Coll. Interface Sci.* **58**, 471.

Hoult, D. I. (1978). *Prog. NMR Spectrosc.* **12**, 41.

Hoult, D. I., and Lauterbur, P. C. (1979). *J. Magn. Reson.* **34**, 425.

Hoult, D. I., Busby, S. T. W., Gadian, R. G., Richards, R. E., and Seeley, P. J. (1974). *Nature* **252**, 285–287.

Kalkwarf, D. R., and Langford, J. C. (1979). Response of agarose solutions to magnetic fields. In "Biological Effects of Extreme Low Frequency Electromagnetic Fields." (R. D. Phillips and M. F. Gillis, eds.) *Conf. 781016*, 408. U.S. Department of Energy.

Kaufman, L., Crooks, L. E., and Margulis, A. R. (1981). "Nuclear Magnetic Resonance Imaging in Medicine." Igaku-Shoin, New York and Tokyo.

Ketchen, E. E., Porter, W. E., and Bolton, N. E. (1978). The biological effects of magnetic fields on man. *Am. Ind. Hyg. Assoc.* **39**, 1.

Kolin, A. (1952). Improved apparatus and techniques for electromagnetic determination of blood. *Rev. Sci. Instr.* **23**, 235.

Lauterbur, P. C. (1973). *Nature* **242**, 190.

Lenz, G. W., Haacke, E. M., Masaryk, T. J., and Laub, G. (1988). In-plane vascular imaging: pulse sequence design and strategy. *Radiology* **166**, 875–882.

Leusden, F. P. (1929). Electric and magnetic effects on bacteria. Zentralbl. Bakteriol. *Parasitenkde Infektionskr.* **111**, 321.

Malmstrom, V. H. (1976). Knowledge of magnetism in pre-Columbian Mesoamerica. *Nature* **259**, 390.

Mansfield, P. M., and Grannell, P. K. (1973). *J. Phys. Chem.* **6**, L422.

Mansfield, P. M., and Morris, P. G. (1982). "NMR Imaging in Biomedicine." Academic Press, New York.

Maret, G., and Dransfield, K. (1977). Macromolecules and Membranes in high magnetic fields. *Physica* **86-88B**, 1077.

Marchal, G., Bosmans, H., Van Fraeyenhoven, L. (1990). Intracranial vascular lesions: optimization and clinical evaluation of three-dimensional time-of-flight MR angiography. *Radiology* **175**, 433–448.

Moon, R. B., and Richards, J. H. (1973). *J. Biol. Chem.* **248**, 7276–7278.

Morgan, C. J., and Hendee, W. R. (1984). "Introduction to Magnetic Resonance Imaging." Multi-Media, Denver.

Murayama, M. (1965). Orientation of sickled erythrocytes in a magnetic field. *Nature* **206**, 420.

Needham, J. (1962). "Science and Civilization in China, Vol. 4 (Part 1)." Cambridge University Press, London.

Neugebauer, D. C., Blaurock, A. E., and Worcester, D. L. (1977). Magnetic orientation of purple membranes demonstrated by optical measurements and neutron scattering. *FEBS Lett.* **78**, 31.

Odeblad, E., Lindstrom, G. (1955). Some preliminary observations on the NMR in biological samples. *Acta Radiologica* **43**, 469–476.

Overhauser, A. W. (1955). *Phys. Rev.* **92**, 411.

Partain, C. L., Price, R. R., Patton, J. A., Kulkarni, M. V., and James, A. E., Jr. (1988). "Magnetic Resonance Imaging, 2nd ed." Saunders, Philadelphia.

Persson, B. R. R., and Stahlberg, F. (1989). "Health and Safety of Clinical NMR Examinations." CRC, Boca Raton, Florida.

Peterson, F., and Kennelly, A. E. (1892). Some physiological experiments with magnets at the Edison Laboratory. *N.Y. Med. J.* **56**, 729.

Polk, C., and Postow E. (1986). "Handbook of Biological Effects of Electromagnetic Fields." CRC, Boca Raton, Florida.

Purcell, E. M., Torrey, H. C., and Pound, R. V. (1946). *Phy. Rev.* **69**, 37–38.

Riederer, S. J. (1988). *Proc. IEEE* **76**, 1095.

Roberts, G. K. M., Wade-Jardetzky, N., Jardetzky, O. (1981). *Biochemistry* **20**, 5389–5394.

Rockette, H. E., and Areana, V. C. (1983). Mortality studies of aluminium reduction plant workers: potroom and carbon department. *J. Occup. Med.* **25**, 549.

Smith, Michael B. (1991). Unpublished results.

Stoner, E. C. (1972). Magnetism. *In* "Encyclopedia Britannica, Vol. 14." Benton, Chicago.

Sperber, D., Dransfeld, K., Maret, G., and Weisenseel, M. H. (1981). Oriented growth of pollen tubes in strong magnetic fields. *Naturwissenschaften* **68**, 40.

Spritzer, C. E., Pelc, N. J., and Lee, J. N. (1990). Rapid MR imaging of blood flow with a phase sensitive, limited flip angle, gradient recalled pulse sequence: preliminary experience. *Radiology* **176**, 255–262.

Swenberg, C. E., and Conklin, J. J. (1988). "Imaging Techniques in Biology and Medicine." Academic Press, San Diego.

Taylor, D. J., Bore, P. J., Styles, P., Gadian, D. G., and Radda, G. K. (1983). *Mol. Biol. Med.* **1**, 77.

Torbert, J., and Maret G. (1979). Fibers of highly oriented Pfl bacteriophage produced in a strong magnetic field. *J. Mol. Biol.* **134**, 843.

Vyalov, A. M. (1971). Clinico-hygienic and experimental data on the effects of magnetic fields under industrial conditions. *In* "Influence of Magnetic Fields on Biological Objects." (Y. Kyaoloov, ed.) National Technical Informational Service, Rep. JPRS5303S, Springfield, Virginia.

Webb, S. (1988). "Physics of Medical Imaging." Adam Hilger, Bristol and Philadelphia.

Wehrli, F. W., Shimakawa, A., Gullberg, G. T., and MacFall, J. R. (1986). Time-of-flight MR selective saturation recovery with gradient refocusing. *Radiology* **160**, 781–785.

Wells, P. N. T. (1982). "Scientific Basis of Medical Imaging." Churchill Livingstone, Edinburg and New York.

Wolff, S. D., and Balaban, R. S. (1988). Magnetization transfer contrast (MTC) and tissue water proton relaxation *in vivo*. *Magn. Reson. Med.* **10**, 135–144.

Williamson, S. J., and Kaufman, L. (1980) Magnetic fields of the human brain. *J. Magn. Mgn. Mater.* 15–18, 1548.

# Index